COAL LIQUEFACTION

Academic Press Rapid Manuscript Reproduction

COAL LIQUEFACTION

The Chemistry and Technology of Thermal Processes

D. Duayne Whitehurst
Thomas O. Mitchell
Malvina Farcasiu

With the Assistance of
Nancy H. Lin

Mobil Research and Development Corporation
Central Research Division
Princeton, New Jersey

ACADEMIC PRESS   1980
A Subsidiary of Harcourt Brace Jovanovich, Publishers

New York   London   Toronto   Sydney   San Francisco

6414-1822 ✓

CHEMISTRY

ACADEMIC PRESS, INC.
111 Fifth Avenue, New York, New York 10003

United Kingdom Edition published by
ACADEMIC PRESS, INC. (LONDON) LTD.
24/28 Oval Road, London NW1 7DX

LIBRARY OF CONGRESS CATALOG CARD NUMBER: 80-19648

ISBN 0-12-747080-8

PRINTED IN THE UNITED STATES OF AMERICA

80 81 82 83   9 8 7 6 5 4 3 2 1

To Heinz Heinemann who suggested the writing of this book.

Contents

CHAPTER 1

INTRODUCTION

CHAPTER 2

THE COMPOSITION OF COAL

CHAPTER 3

COAL LIQUIDS

CHAPTER 4

SIGNIFICANCE OF THE PHYSICAL PROPERTIES OF COAL TO COAL CONVERSIONS

CHAPTER 5

COAL RANK AND LIQUEFACTION

CHAPTER 6

CATALYTIC EFFECTS OF MINERAL MATTER ON COAL LIQUEFACTION

CHAPTER 7

EFFECTS OF PROCESS VARIABLES (TIME, TEMPERATURE, AND HYDROGEN GAS PRESSURE) IN SRC PRODUCTION FROM COAL

Foreword

Fossil fuels were created by the gradual conversion of organic materials and residues over millions of years. Our society has used them as sources of energy. We have broadly differentiated between three major classes, based on physical state and use: natural gas, petroleum, and coal. They are all members of a chemically related family of fossil resources that include yet other materials such as tar sands, oil shales, heavy oils, etc. All fossil fuels have carbon skeletons with varying amounts of hydrogen atoms attached, and functional groups containing heteroatoms (mainly oxygen, sulfur, and nitrogen).

For many years the petroleum chemist and technologist have studied and dealt with a broad spectrum of such materials in "petroleum" itself: from the nearly heteroatom-free hydrocarbon skeletons in the "light" end of petroleum, to the "residual" fractions of petroleum that consist of semisolids and contain structures and functional groups similar to those in coal.

Our future use of resources will include a shift toward the latter more complex and condensed materials, including the residual petroleum fractions and coal. It was logical therefore to extend and expand our chemical knowledge and skills that developed largely in the context of petroleum science further into the direction of semisolids and solids (coal). The authors have been in the midst of this evolutionary development.

This volume describes the findings and knowledge gained from a cooperative effort conducted by Mobil and the Electric Power Research Institute in a four-year program. This has been a unique research effort: It has combined talents, knowledge, and efforts of both industrial and academic communities. It deals with the generation of fundamental knowledge, yet was objective-oriented toward a common goal. It is a pleasure to acknowledge the broad spectrum of important contributors. These included: researchers and scientists of the Department of Aerospace and Mechanical Sciences at Princeton University; College of Earth and Minerals Sciences and Chemistry Departments at thePennsylvania State University; Chemistry Department, University of California at Berkeley; Chemistry Department, University of Iowa; and theThermochemistry and Chemical Kinetics Group, Stanford Research Institute.

In making the step from a large amount of knowledge and interpretations to a coherent book, the formidable task of satisfying all authors, co-workers, publisher, sponsors, management, etc., was accomplished by N. H. Lin, as editor and coordinator.

P. B. Weisz

Mobil Research and Development Corporation
Central Research Division
Princeton, New Jersey

Preface

This book describes the conversion of coal to liquids in terms of the structural and functional differences between the coal and the desired liquid products. It should be of general interest to libraries, academic institutions, and individuals as a general reference on coal chemistry. Chemists and engineers just entering the field of synthetic fuels from coal will find this book particularly useful in developing an understanding of the chemistry involved in coal transformation. It could also serve as a supplemental text in college courses in alternative energy sources.

Although this book is primarily devoted to thermal processes for coal liquefaction, it also provides much insight into catalytic liquefaction processes through discussion of the fundamental chemistry of coal.

A direct link is shown between the fundamental chemistry and composition of coal (and the transformations that are involved in thermal coal liquefaction) and the approaches used in the currently developing technology. There is a logical sequence of presentation in which the composition of coal is reviewed, followed by a description of the composition of liquid coal products. The significance of the physical properties and rank of coal relative to its conversion are then discussed. The more subtle aspects of coal conversion such as the catalytic effects of mineral matter, effects of process parameter variation, problems with side reactions, such as char formation, and the relationship between solvent composition and liquefaction behavior are then brought into focus. Finally, these initial discussions are tied together, showing the intrinsic limitations in present-day processes, and some suggestions are made for possible improvements.

This book is unique in its approach and should provide the reader with a more fundamental understanding of the limitations in coal liquefaction from a chemical viewpoint than is found in other texts.

D. D. Whitehurst

Principal Investigator

Acknowledgments

In concluding the research that made this book possible, a number of individuals and organizations contributed a great deal to the overall effort. Although it is impossible to list all of the individuals who have influenced this book, we would like to acknowledge the following, who played major roles in obtaining the necessary information. A number of other individuals are acknowledged for their contributions in the text of the book.

W. C. Rovesti, Project Manager of EPRI that cofunded the work with Mobil Research and Development Corporation.

E. Plett, F. Rodgers, and A. Alkidas of Princeton University, Department of Aerospace and Mechanics, who helped study the effects of coal particle size, pretreatment, and stirring on coal conversion, conducted some very short reaction time conversions, and performed coulometric analyses of coal liquids.

A. Pines and D. Wimmer of the University of California at Berkeley, Department of Chemistry, who conducted solid state ^{13}C-NMR analyses of coals, SRCs, and residues.

A. Davis, G. D. Mitchell, and R. G. Jenkins of the Pennsylvania State University, College of Earth and Minerals Sciences, who conducted petrographic and optical microscopic analyses of coals and residues as well as studies on the mechanisms of char formation during coal liquefaction.

M. Anbar and S. E. Buttrill, Jr., of Stanford Research Institute, who characterized SRCs and solvents by field ionization mass spectroscopy.

S. Wawzonek of the University of Iowa, Department of Chemistry, who conducted polarographic analyses of coal liquids.

We also wish to express our appreciation to N. H. Lin, who spent many long hours improving the quality of the text, to J. C. Zahner for his many helpful comments on early drafts, and to L. M. Cole, L. Kardish, and L. Olivacz without whose secretarial help this book would not have been possible.

Chapter 1

INTRODUCTION

Interest in converting coal into liquid products has been
rather cyclic and invariably controlled by access to petroleum.
In the beginning of the industrial revolution (∿1840), coal was
the major source of energy in the United States as well as in the
rest of the world. Coal liquids were byproducts in the manufacture
of coke needed for the growing steel industry and were used primarily as chemical feedstocks. Hydrogenative coal liquefaction
had its beginning in 1869 when Bertholet reported that coal could
be converted by reduction to an oil-like product (1-1).

Coal continued to dominate the United States' energy supply
for the next hundred years. Figure 1-1 shows the trend of energy
consumption in the United States from the beginning of the industrial revolution to 1978 (1-2,1-3). The figure shows how quickly
petroleum became the preferred energy source after its discovery
in 1859 in Pennsylvania and its rapid commercial production in the
early 1900's.

By the early 1920's, worries that oil supplies were being
depleted caused research in coal liquefaction to flourish. This
interest was accentuated by an expanding automobile industry.
However, the subsequent discovery of oil in Texas in the mid-1920's
discouraged further work in the United States. Other countries
which were not so well endowed with indigenous oil resources continued to seek methods for converting coal to more convenient
liquid fuels. In particular, Germany laid the foundations for most
of the present day direct and indirect coal liquefaction processes
with the discovery and development of the Bergius hydrogen donor
process in 1913, the Pott-Broche solvent extraction process in
1927, and the Fischer-Tropsch process for hydrocarbon synthesis
from gasified coal in 1925 (1-4).

During World War II, petroleum shortages in Germany led to the
construction of 27 synthetic fuels plants with a maximum total

1

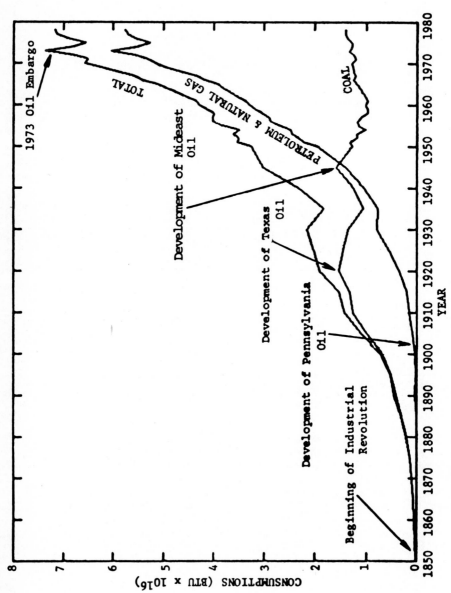

FIGURE 1-1. Consumption of Fossil Fuels in the United States

output of 5 million tons/year ($\sim$86,000 bbl/day) (1-5). These liquid fuels supplied all of Germany's aviation fuel and $\sim$75% of her liquid fuel needs during the war.

To put this in perspective, 5 million tons/year is about the same consumption level as petroleum in the United States in 1900.

After World War II, the United States experienced petroleum shortages as the rate of production could not keep up with the growing demand. Coal liquefaction was again considered as an alternative and a sizeable research effort was initiated. Several commercial demonstration plants (up to 300 tons/day) were built and operated for several years (1-4,1-6). This interest, however, was short-lived as the discovery of massive petroleum reserves in the Middle East in the mid-1940's once again made coal liquefaction uneconomical. From that point on, the level of effort declined until 1973 when unrest in the Middle East made the United States keenly aware of her vulnerability because of the dependence on foreign countries for her major source of energy. At the same time, the United States' crude production began to decline.

Recent statistical surveys indicate that more than half of all the possible petroleum reserves of the United States (including Alaska) were consumed by 1979 (1-7). Thus, because of continual petroleum depletion, uncertain supplies of foreign crude, and rapidly increasing costs of imported petroleum, coal liquefaction is again being considered as a major source of liquid fuels.

It was in light of these developments that the work described in this book was begun in 1975. Although many processes had been developed prior to the recent interest in coal liquefaction, little was known of the composition of coal or the detailed chemistry involved in transforming this insoluble material into the quality fuels that have now become standard.

The structure of coal is very different from that of hydrocarbons found in petroleum. Processes which attempt to use technology developed for petroleum upgrading may not be optimal for coal processing; thus transformations specific for coal must be considered. In particular, the high content of heteroatoms and the low content of hydrogen make the initial coal products very susceptible to the formation of coke and gas. Such side reactions can cause serious technological problems and lower the overall thermal efficiency. Efficient hydrogen gas consumption is the major key to developing improved processes.

The majority of the text presented in this book is the result of a concerted effort aimed at answering fundamental questions about coal structure, the mechanisms involved in its conversion to liquid fuels and the intrinsic limitations of coal transformation to liquids. The project from which it originated was entitled "The Nature and Origin of Asphaltenes in Processed Coals," and was co-sponsored by the Electric Power Research Institute (EPRI) and Mobil Corporation (RP-410-1). Unless otherwise referenced, data and results reported in this book are from that work.

The bulk of the work was carried out in the Central Research
Division Laboratory of Mobil Research and Development Corporation
(MRDC) in Princeton, New Jersey. D. D. Whitehurst was the principal investigator. M. Farcasiu was responsible for the compositional characterization studies of coal and coal liquids, and
developed several liquid chromatographic techniques which were
keys to obtaining fundamental understanding of coal and coal
liquid structures. T. O. Mitchell developed a "synthetic" recycle
solvent which enabled the simultaneous study of the chemistry of
the solvent and the chemistry of the coal. He also was responsible for the bulk of the coal conversion work which aided in our
understanding of the intrinsic limitations of process conditions
and the mechanisms involved in coal transformations. J.J. Dickert
aided in the conversion studies and was responsible for the studies
relating to the development of an understanding of the effects of
mineral matter on coal conversion. B. O. Heady and G. A. Odoerfer
were also integral parts of the effort and developed much of the
instrumentation used for coal conversion and the characterization
of coal liquid products.

The project was also aided by cooperative efforts with many
other groups which were either supported directly by the project
or indirectly through EPRI. It is impossible to list all of the
individuals who contributed to this effort but the major contributors are summarized in the acknowledgement.

This book presents the results of this work and attempts to
integrate this new information with past as well as current work
at other laboratories. The theme of the book is to show how
detailed knowledge of the structure and chemistry of coal, the
mechanisms involved during its conversion, and the intrinsic
limits of the chemistry which functions at the high temperatures
needed for coal conversion are keys to controlling the hydrogen
requirement and can lead to the development of more efficient
liquefaction processes.

REFERENCES

1-1. M. Bertholet, <u>Bull. Soc. Chem. Fr.</u> <u>11</u>, 278 (1869).
1-2. J. C. Fisher, "Energy Crises in Perspective", John Wiley and
      Sons, Inc., New York, 1974, p. 157.
1-3. Energy Statistics, Vol. 2, No. 3, Third Quarter 1979, Insti-
      tute of Gas Technology, Chicago, Illinois.
1-4. "Liquid Fuels from Coal", published by National Coal Board,
      United Kingdom, August 1978, p. 19.
1-5. "Report on the Petroleum and Synthetic Oil Industry of
      Germany", Ministry of Fuel and Power, Great Britain, MHSO,
      London, 1947.
1-6. I. Howard-Smith and G. J. Werner, "Coal Conversion Technol-
      ogy", Noyes Data Corporation, Park Ridge, N.J., 1976.
1-7. M. K. Hubbert, "World Resources of Fossil Organic Raw
      Materials", taken from Resources of Organic Matter for the
      Future, Toronto 1978 (Editor of Reprints, Dr. L.E. St-Pierre),
      Multiscience Publications Limited, Montreal, Quebec, Canada,
      p. 58.
1-8. D. D. Whitehurst, M. Farcasiu and T. O. Mitchell, "The
      Nature and Origin of Asphaltenes in Processed Coals",
      EPRI Report AF-252, First Annual Report Under Project
      RP-410, February 1976.
1-9. D. D. Whitehurst, M. Farcasiu, T. O. Mitchell and J. J.
      Dickert, Jr., "The Nature and Origin of Asphaltenes in
      Processed Coals", EPRI Report AF-480, Second Annual
      Report Under Project RP-410-1, July 1977.
1-10. D. D. Whitehurst, T. O. Mitchell, M. Farcasiu and J. J.
      Dickert, Jr., "The Nature and Origin of Asphaltenes in
      Processed Coals", Final Report Under Project RP-410-1,
      March 1977-January 1979, in preparation.

Chapter 2

THE COMPOSITION OF COAL

I. INTRODUCTION

Coal is believed to originate predominantly from plants. In fact, individual plant constituents are recognizable in today's coal as discrete fossilized plant fragments called macerals (which identify coal as we know it today). Figure 2-1 shows the major macerals commonly identified and their plant origins.

The original plant matter, deposited in sedimentary strata, was subject to bacterial action, other chemical alterations and geological effects, such as compaction. The decaying matter was systematically transformed through a series of evolutionary changes. The first product was humic acid, which then converted sequentially into peat, subbituminous coal, bituminous coal, and finally into anthracite. With these changes, the carbon content increased and the oxygen content decreased, resulting in higher calorific values for higher ranked coals, as also shown in Figure 2-1.

The plant constituents which could possibly give rise to coal are shown in Figure 2-2. Although these structures are present in plants today, and could be similar to those of plants of prehistoric times, it is not anticipated that they would survive intact over the long periods of time required for their transformation to coal. Some of the structural features, however, may be recognized even in today's coal.

United States' coals usually consist of at least 80% vitrinite but there are differing opinions on whether the composition of this vitrinite is the result of the coalification of cellulose or of lignin structures. Given and others have shown that in plants which are decomposing today, the cellulose undergoes very rapid biodegradation (2-2) -- perhaps too rapidly for it to be a major coal precursor. Protein also degrades rapidly. The plant constituents which are most resistant to bacterial attack are lignins, waxes,

resins, tannins, flavonoids, and possibly alkaloids (2-3). Lignin
structures or "woody tissue" may, therefore, constitute the majority of the decayed plant components (2-1). Given's work has also
shown that certain components of coal can be related to structures
evolved from lignins (2-4).

As a point of reference, it should also be remembered that the
U.S. coals were laid down in two different geological ages, about
160 million years apart. The Carboniferous period produced Eastern
coals and the Cretaceous period later yielded Western coals. The
structures associated with these two geological ages may be substantially different.

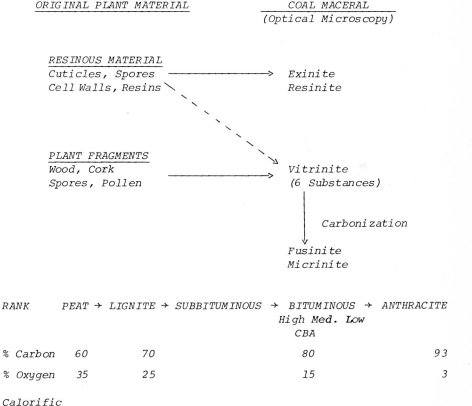

RANK	PEAT →	LIGNITE →	SUBBITUMINOUS →	BITUMINOUS →			ANTHRACITE
				High Med. Low			
				CBA			
% Carbon	60	70		80			93
% Oxygen	35	25		15			3
Calorific Value (BUT/# MAF)	12000	13000		14000	16000		15500

FIGURE 2-1. Mode of Formation of Coal

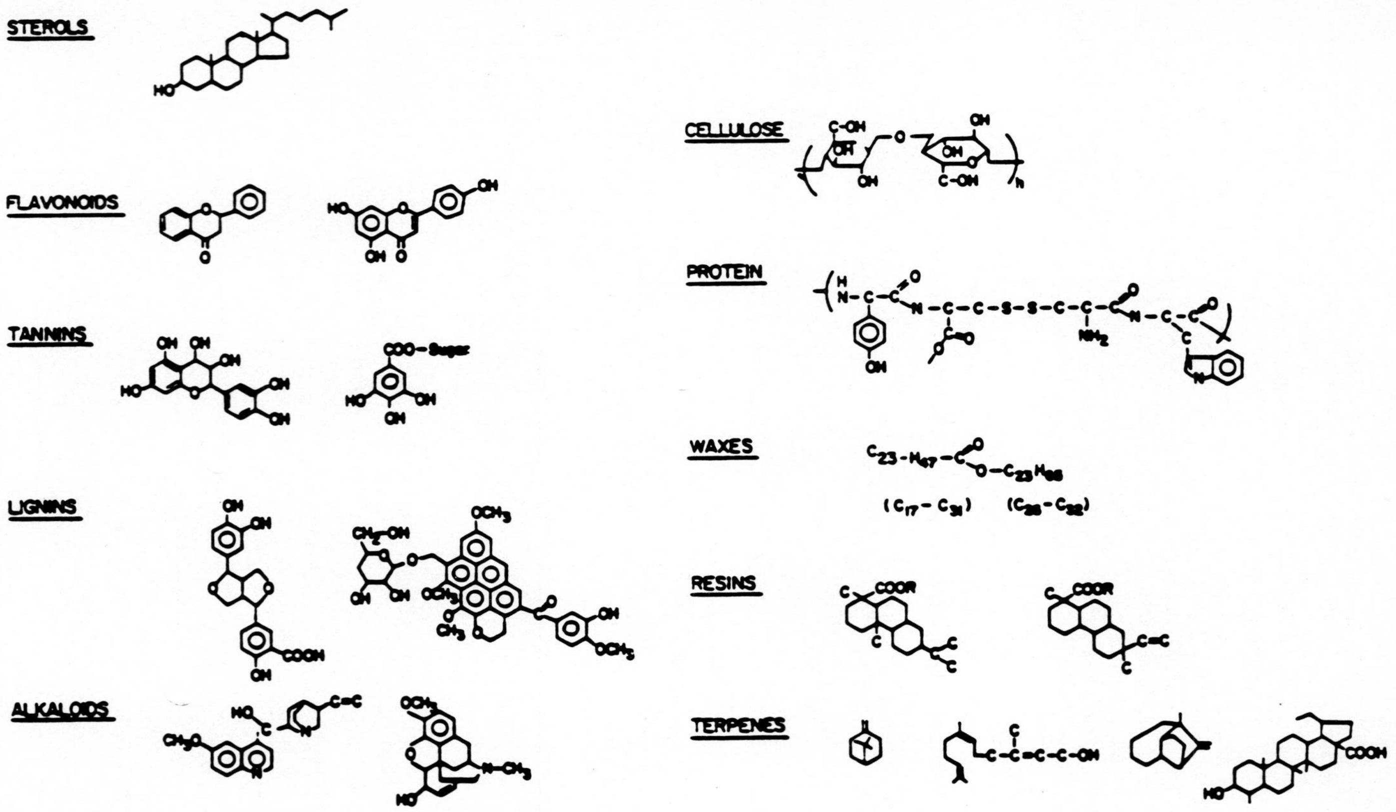

CELLULOSE

PROTEIN

WAXES

RESINS

TERPENES

STEROLS

FLAVONOIDS

TANNINS

LIGNINS

ALKALOIDS

FIGURE 2-2. Structures of Coal Precursers

II. AROMATICITY OF COAL

There is controversy on the proposed primary backbone structure of coal. Some workers contend that coal is primarily graphite-like; others argue that coal is of a diamond-like structure. Both of these structures are low in H/C ratio which is consistent with the composition of coals.

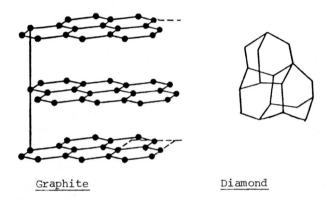

Graphite   Diamond

To gain insight on the structure of coal, past workers have attempted to break coal down into recognizable units and then piece them back together as is done in natural product chemistry today. The most common technique presently pursued is that of oxidative degradation. With oxidants such as HNO_3, $K_2Cr_2O_7/HNO_3$, $KMNO_4/OH-$, BuOOH/AIBN, or peracetic acid, workers have come to the conclusion that coal is predominantly aromatic and contains many condensed rings (2-4 - 2-6). Other authors using NaOCl/OH- have come to a different conclusion; they believe coal contains large amounts of quarternary aliphatic carbon, or is diamond-like in structure and contains 50% aromatic carbon or less (2-7).

The preceding methods of oxidation selectively oxidize only the aliphatic portion of coal. A new method pioneered by Deno (2-8) uses trifluoroacetic acid in combination with hydrogen peroxide, and selectively oxidizes the aromatic rings. These two techniques could be combined into a powerful tool for structural characterization of coal.

Because of the difficulties in piecing together the frag-mented products of coal, a number of workers have attempted direct characterization of coal. Direct techniques, however, suffer because coal is an opaque solid which is essentially insoluble in its natural form, and relatively few tools have been available for such direct measurements. In the past, techniques such as X-ray scattering have been used and conflicting interpretations as to

the dominant structure of coal have been reported. Hirsch first
reported that coal was from 50-80% aromatic, with primarily 89%
ordered structure (2-9). Later Ergun, using X-ray scattering,
concluded that coal is less aromatic and contains large quantities
of amorphous regions (2-10). Friedel, using ultraviolet techniques, concluded that coal could not be polyaromatic and contained large amounts of aliphatic structure (2-11). Given, in characterizing coal extracts by polarographic reduction concluded that
low-rank coals were greater than 20% aromatic and high-rank coals
were greater than 50% aromatic (2-12). Polycyclic aromatic rings
were believed to be prevalent. Recently, new tools have evolved
and for the first time coal can be characterized directly in its
natural form. The most promising of these tools are a solid state
CP-^{13}C-NMR technique developed by Pines (2-13) and Fourier Transform Infrared Spectroscopy (which can provide quantitative IR
spectra on solid materials).

    Working in conjunction with Pines, we have found that there
is relatively little correlation between the hydrogen/carbon mole
ratio and the percent aromatic carbon found in coal or coal
liquids, as shown in Figure 2-3. These data give some indication

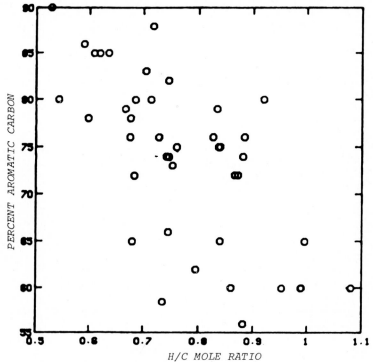

FIGURE 2-3. Aromatic Carbon vs. H/C

of why correlating aromatic carbon content with the elemental composition of the coal has been so difficult. There is, however,
some correlation between the rank of the coal and its aromaticity
(Figure 2-4). The aromatic carbon content increases from about
50-60% for some subbituminous coals to over 90% for anthracite.
This aromaticity changes with coal conversion as will be shown
later. The CP-^{13}C-NMR technique is somewhat new and still evolv-
ing. It shows promise in characterizing coal into aliphatic and
aromatic components and further subdividing the structural types.
As representative examples of the spectra that one can achieve,
Figure 2-5 shows typical model compounds, viz p-diethoxybenzene
and 2,3-dimethylnaphthalene, a parent coal, SRC derived from this
coal, unconverted residue and the spherical coke formed on extended
thermal reaction. It is possible to distinguish between aromatic,
aliphatic, aliphatic ether, and condensed aromatic carbon in the
model compounds.

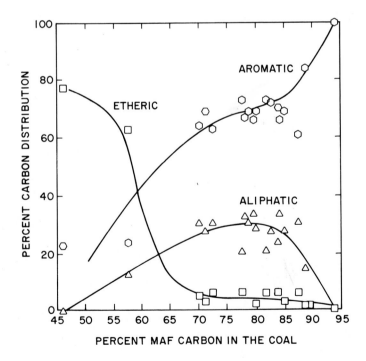

FIGURE 2-4. Carbon Distribution for Coals of Varying Rank

D. Duayne Whitehurst *et al.*

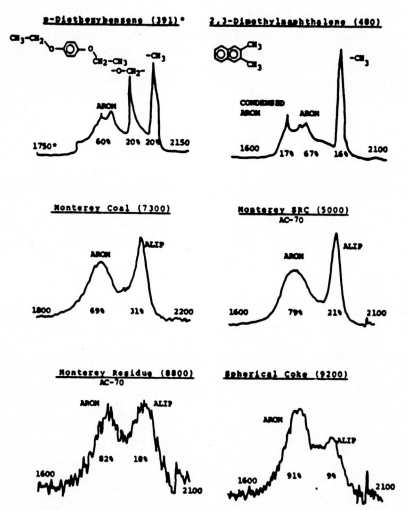

FIGURE 2-5. CP-^{13}C-NMR Spectra of Representative Samples.
The numbers in parentheses indicate the number of spectra accumulated to generate the spectrum presented. Note that the scales are not all the same.

Organic polarography is another valuable tool in evaluating the nature of the aromatic ring system. This method is described in detail later but we shall mention here that from the initial products of thermal coal conversions, coal is inferred to contain an average of only 2 or 3 fused aromatic rings rather than large fused ring systems per molecular unit.

To put coal in perspective with conventional liquid fuels, one can compare H/C ratios. Figure 2-6 shows that coal must undergo major changes in H/C mole ratio if it is to be transformed into synthetic fuels which resemble the liquid fuels to which we have become accustomed.

III. FUNCTIONAL GROUPS IN COAL

The predominant functional groups found in coal are those containing oxygen. Specifically, they include phenols and alcohols, ethers, carboxylic acids and carbonyls. In view of the diversity of decomposed plant matter, finding any systematic variation in the distribution of these functional groups seems unlikely. However, analyses of coals varying in rank have shown that rather surprising correlations exist between the MAF carbon content of coals and the content of various functional groups (2-14 - 2-19).

Quantitative determination of these various oxygen functionalities has been reviewed by several authors (2-14, 2-15, 2-19) and specific methods of analysis have been carried out on a variety of coals. For the purpose of this book, we have chosen for discussion those analyses which we feel are the most reliable to date. Data for phenols were taken from Given (2-16), carboxylic acids from Schafer (2-17), and methoxy and carbonyl groups from van Krevelen (2-14). The ether contents were calculated by difference and would therefore include any other undetermined oxygen functionality.

The systematic variation of these functional groups with the carbon content of the coal is shown in Figures 2-7(a) - 2-7(d). Also shown in these figures is a fitted line (first or second order) based on the best predictive equation for each functionality.

The values of coals varying in rank from 55 to 95% MAF carbon were determined from these equations and are shown in Figure 2-8 as the cumulative percent oxygen found in different functional groups for the different coals. It can be seen that methoxy and carboxy functionalities are of little importance at greater than 80% carbon. Moreover, if one calculates the relative molar concentrations of these functional groups and integrates what is

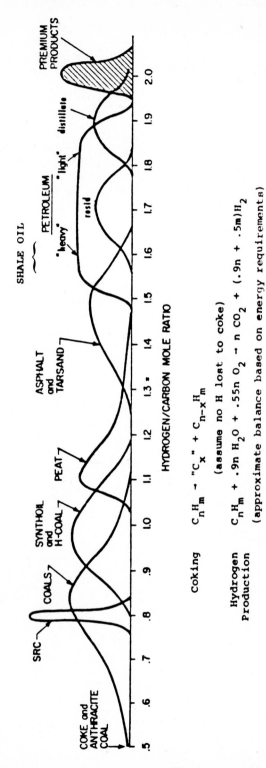

Coking $C_nH_m \rightarrow "C_x" + C_{n-x}H_m$

(assume no H lost to coke)

Hydrogen $C_nH_m + .9n\ H_2O + .55n\ O_2 \rightarrow n\ CO_2 + (.9n + .5m)H_2$
Production

(approximate balance based on energy requirements)

FIGURE 2-6. H/C Ratios for Various Hydrocarbon Sources

14

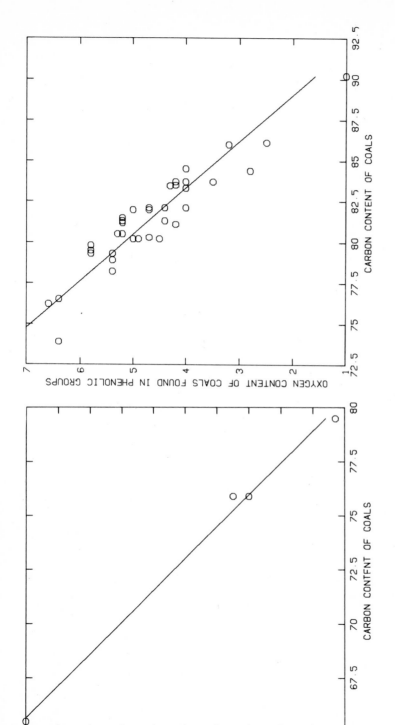

FIGURE 2-7 (b). Hydroxyl (Phenol) Content of Coals

FIGURE 2-7(a). Methoxl Content of Coals

15

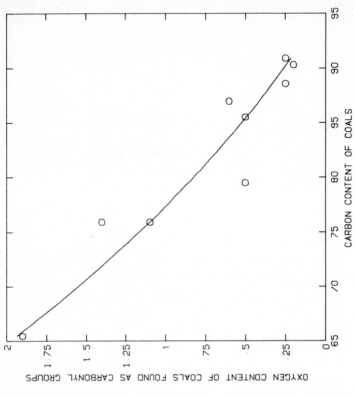

FIGURE 2-7(c). Carboxylic Acid Content of Coals

FIGURE 2-7(d). Carbonyl Content of Coals

known of the aromatic content and type in the coal, a general picture of the composition of coals of various rank can be generated.
For example, coal conversions often yield products with molecular
weights from 300 to 600, and it appears that rather stable molecular units of these molecular weights exist in coals. A 30-carbon
fragment would be representative of an average molecule of this
molecular weight. Polarographic analysis of the initial products
of coal dissolution show little evidence for condensed aromatic
nuclei containing more than 2 or 3 condensed rings. For the following discussion, aromatic nuclei are assumed to be represented as
naphthalene units (or 10-carbon groupings). Using this unit and
the above data, the number of aromatic rings and functional groups
per average molecular unit (30 carbons) can be calculated. Such
calculations can aid in understanding the differences between coals
in terms of their chemistry. Table 2-1 shows typical compositions
of a lignite, a subbituminous, and a bituminous coal.

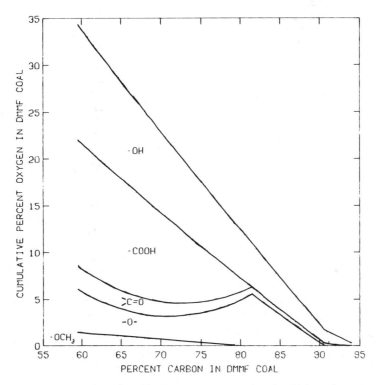

FIGURE 2-8. Distribution of Oxygen Functionality in Coals

TABLE 2-1. Average Composition of Coals Varying in Rank

Coal	% C	No. of 10 C Aromatic Rings	Mol Wt.	Empirical Formula				# Functional Groups			
				C	H	O	N	COOH	OH	$-O-$ [a]	$\overset{O}{\underset{C}{\parallel}}$ OCH$_3$
Lignite	70.0	1.7	513	30	28	7.4	0.5	1.5	2.8	0.8	0.5 0.2
Subbituminous	77.5	2.0	461	30	24	4.4	0.5	0.6	1.8	1.1	0.3 0.1
Bituminous	85.0	2.3	419	30	22	1.9	0.4	0	0.9	0.9	0.1 0

[a]Ether or unaccounted-for oxygen.

The bituminous coal would thus contain no carboxylic acids or methoxy groups, one ether, one phenol, and slightly more than two aromatic nuclei per molecular unit. The subbituminous coal would have about one carboxylic acid, one ether, two phenols, and two aromatic nuclei. The lignite would contain more than one carboxylic acid, three phenols, one ether and less than two aromatic nuclei. In later chapters we shall discuss how these functional groups are affected during coal liquefaction.

Several excellent reviews are available on the forms of sulfur found in coal (2-20 - 2-22). Two major classes predominate, inorganic sulfur (mostly pyrite) and organic sulfur. The original decaying plant matter contained up to 0.5% sulfur. Any higher sulfur contents were caused by sulfate-reducing bacteria (2-20 - 2-22) as illustrated in Figure 2-9. The organic matter was submerged in saline or brackish water containing sulfate salts. The bacteria consumed a portion of the organic matter and used SO_4 as an oxygen source, thus producing a variety of reduced-sulfur containing ions (HS-, HSX, etc.). If iron ions were present in the water, iron sulfide precipitated and produced pyrite. The coal's organic portion also changed as the oxygen functionality was exchanged for sulfur.

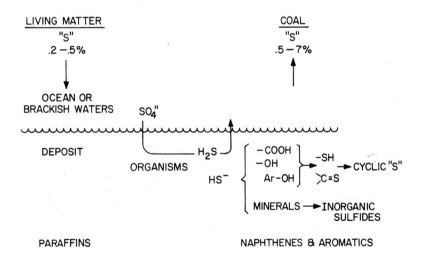

FIGURE 2-9. Origin of Organic Sulfur in Coal

The kinetics of pyrite and organic sulfur production are therefore parallel. The ratio of organic to inorganic sulfur will thus
be unique to a given location as dictated by the composition of
the dissolved salts in the deposit and physical parameters such as
time, temperature, and pressure.

Because sulfur is introduced via exchange of oxygen functionality, the chemistry of the sulfur species is similar to that of
the oxygen species in the coal. Major functional groups are thus
thiols, sulfides, and heterocyclic aromatic rings (thiophene derivatives).

Sulfoxides are known to be present in tar sands, which have a
similar mode of sulfur incorporation (2-20). Their presence in
coal is less well defined.

In contrast to sulfur, nitrogen functionality in coal is believed to be representative of the nitrogen species of the original plant matter. This explains the rather low degree of variation
in the nitrogen content of all coals (1-2%). The predominant functional groups are heterocyclic rings such as pyridine or pyrrole
derivatives. Metals are found as salts or associated with clays
or porphyrins.

IV. SPECULATIONS ON COAL STRUCTURE

With the data described above, one can proceed further and
even speculate on an average "molecular structure" of a coal. A
number of workers have attempted to derive such representative
structures for bituminous coal which are consistent with the observed chemistry. One of the first structures was by Given, and
is shown in Figure 2-10. This structure was not intended to be
"the" structure for coal but merely to represent the kinds of
structures one should envision as constituting coal (2-24).

This structure is consistent with the coal's elemental composition, with highly substituted but not highly condensed aromatics,
and with functionalities known to be present in coal. A more recent, more sophisticated model was presented by Wiser (2-25) and
is given in Figure 2-11. The significance of this figure is the
location of a number of relatively weak bonds indicated by arrows
which can account for the rapid breakup of coal into smaller more
soluble fragments during coal liquefaction.

In view of the chemistry of the solvent refining of coal and
the composition and chemical nature of the soluble products (to be
discussed in detail later), it is possible to envision the original structure of coal in a more gross sense as a highly crosslinked
amorphous polymer, which consists of a number of stable aggregates
connected by relatively weak crosslinks. As shown in Figure 2-12,
at high temperatures, this highly crosslinked structure thermally
fragments into radicals which in the presence of hydrogen donor

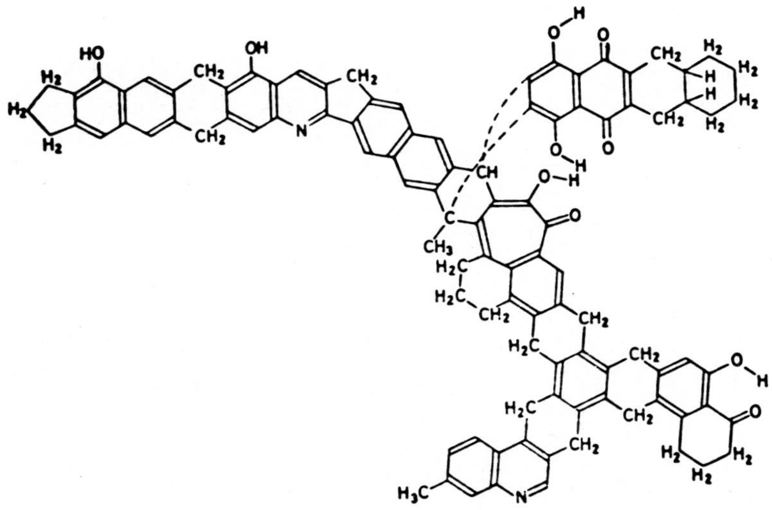

FIGURE 2-10. Proposed Structural Elements of Coal by Given (2-12).

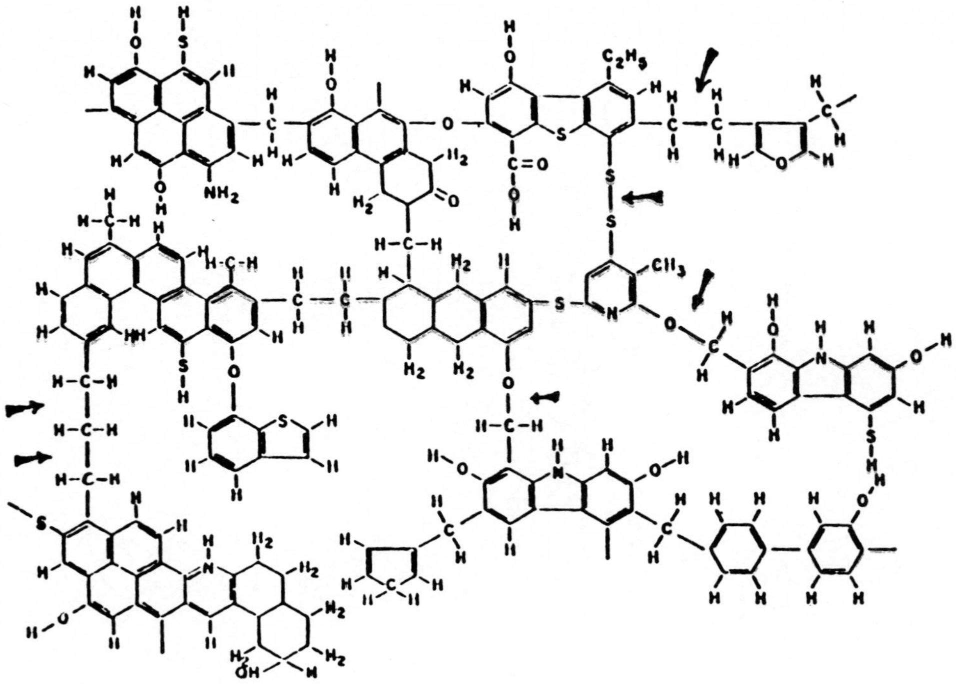

FIGURE 2-11: Representation of Functional Groups in Coal (2-26). Arrows indicate weak bonds.

22

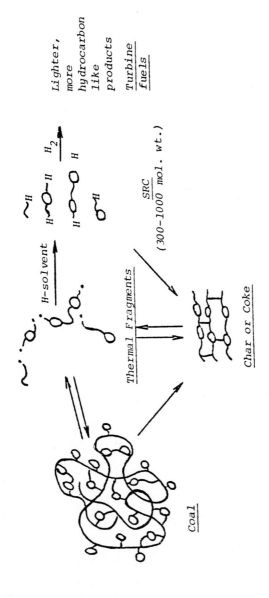

FIGURE 2-12. Chemistry Involved in Coal Conversion

23

COAL RANK COMPOSITION	Kentucky Bituminous $C_{100}H_{82}N_{1.7}O_{9.4}$	Hiawatha Bituminous $C_{100}H_{84}N_2O_{9.5}$	Monterey Bituminous $C_{100}H_{88}N_{1.6}O_{13.2}S_{1.74}$	Nyodak Subbituminous $C_{100}H_{86}NO_{18}$
# Polycondensed Saturated Rings/100 Carbons — Coal / SESC-3 / SCT[a]	4 / 2		6 / 4	10 / 5
Estimated Average Structure SESC-3 SCT-SRC	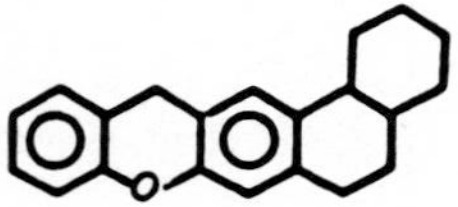	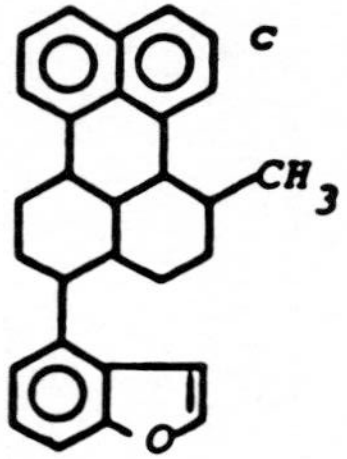 [c] CH_3	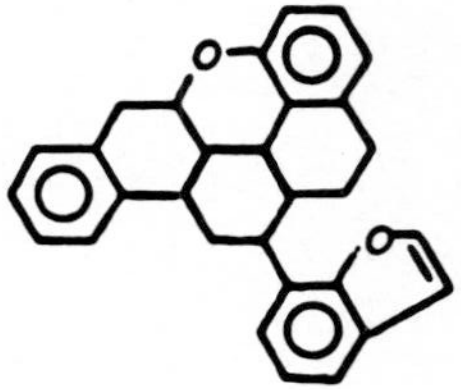	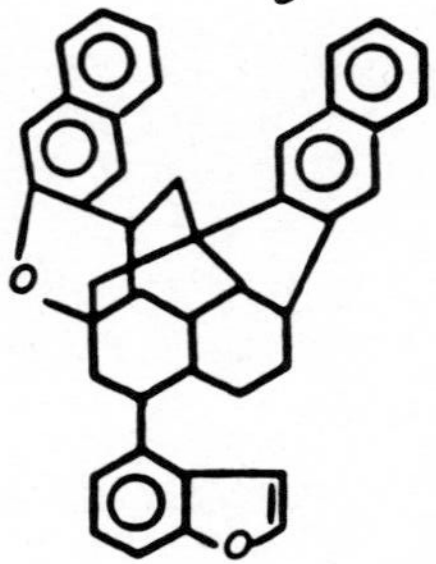
g H Consumed/100g Coal During Liquefaction (Normalized to 100% Conversion) — SCT / LCT[b]	0.43 / 0.96		0.63 / 2.03	0.89 / 2.80

[a] SCT = Short Contact Time <2 minutes.
[b] LCT = Long Contact Time >30 minutes.
[c] Ca. 30 minutes contact time.

FIGURE 2-13. Comparison of Structures and Hydrogen Consumption of Four Coals

solvents are capped and appear as stable species. These species
consist of aggregates or groups of aggregates and correspond to
molecular weights in the 300 to 1000 molecular weight range. In
the absence of hydrogen donor solvents, the original radicals or
the smaller soluble species may recondense to form char or coke.

A relatively new method for deriving coal structure is to convert coals only to the extent that they can be dissolved in a very
polar solvent such as pyridine (2-26,2-27-2-29). Reaction conditions of 1 to 5 minutes in hydrogen donor solvents at 800°F to
840°F are sufficient to convert more than 80% of the coal and over
90% of the maximum achievable conversion of the coal.

With this technique, the weak bonds are broken and capped by
hydrogen but little change in the skeletal structure occurs. The
aromatic carbon and hydrogen content of the soluble products are
almost identical to that of the parent coal. The functional groups
are only slightly altered; decarboxylation is usually complete but
only some polyaromatic ring phenols are dehydroxylated (2-30). By
detailed characterization of certain fractions of the soluble products, a good picture of the predominant skeletal structures of
the coal can be obtained. The characterization and the derivation
of these structures are described in detail in the following chapter. For ease of discussion, we show in Figure 2-13 the types of
skeletal structures that are found in various coals.

It can be seen that the bituminous coals consist of more planar ring structures, whereas the subbituminous coal contains a
highly condensed aliphatic structure (2-27). It is not known if
this particular structure is predominant among subbituminous coals
or is specific to the subbituminous coal that was examined. Hydrogen peroxide/trifluoroacetic acid oxidation of an Alaskan subbituminous coal produced an almost identical product spectrum to that
of the Wyoming subbituminous coal (2-30). Thus the skeletal structure must not be grossly different (2-31).

V.   SUMMARY

The following statements summarize the presently available information on coal composition:

There is little evidence for highly condensed aromatic rings
in coal or the initial liquefaction products. Mono- and diaromatic rings predominate.

High aromaticity in coal products is the result of the processes being used to convert coal, not an intrinsic property of
the starting material.

The aromaticity of coal varies with rank and can be as low as
about 40% for some subbituminous coals.

Coal structures cannot be described by a continuum of changing skeletal structures.

Subbituminous coals of the Cretaceous Period contain significant amounts of three-dimensional polycyclic aliphatic rings, while the Eastern bituminous coals of the Carboniferous Period contain more planar structures.

Functionality appears to decrease systematically as rank increases.

REFERENCES

2- 1.   W. Francis, "Coal, It's Formation and Composition",
        Edwaard Arnold (Publishers) Ltd., London, 1961.
2- 2.   P. H. Given, Personal Communication.
2- 3.   A. D. Chen, "Flavonoid Pigments in the Red Mangrove,
        Rhizophora mangle L. of the Florida Everglades and in the
        Peat Derived from it", PhD Thesis, The Pennsylvania State
        University, September, 1971.
2- 4.   P. H. Given, J. Bimer, S. Raj., Oxidative Study of the
        Structure of Vitrinites, presented at the Fuel Division
        Chicago ACS Meeting, August 1977.
2- 5.   Reference 1, pp. 717-755.
2- 6.   R. E. Winans, R. Hayatsu, R. G. Scott, L. P. Moore and M. H.
        Studier, "Examination and Comparison of Structure, Lignite,
        Bituminous, and Anthracite Coal", presented at the SRI Coal
        Chemistry Workshop, Menlo Park, California, August 1976.
2- 7.   S. K. Chakrabartty and N. Berkowitz, Fuel (1974) 53, 240.
2- 8.   N. C. Deno, Fuel, in press.
2- 9.   P. B. Hirsch, Proc. Inst. Fuel Conf., "Science in the Use
        of Coal", Sheffield Eng. 1958, p. A-29.
2-10.   S. Ergun and V. Tiensuu, Nature (1959), 183, 1668. Acta
        Cryst. (1959), 12, 1050.
2-11.   R. A. Friedel and J. A. Queiser, Fuel (1959), 38, 369.
2-12.   P. H. Given and M. E. Peover, Fuel (1960), 39, 463.
2-13.   A. Pines, M. G. Gibby and J. S. Waugh, J. Chem. Phys.
        (1973), 59, 569.
2-14.   L. Blom, L. Edelhausen and D. W. vanKrevelin, Fuel, 36,
        (1957).
2-15.   A. Ihnatowicz, Glownego Instytutu Gornictwa Komanikat,
        No. 125, Katowice, 1952.
2-16.   P. H. Given, P. L. Walker, W. Spackman, A. Davis and H. L.
        Lovell, "The Relation of Coal Characteristics to Coal Lique-
        faction Behavior," Report No. 4 from The Pennsylvania State
        University submitted to The National Science Foundation,
        December 1, 1975, p. 13.
2-17.   H. N. S. Schafer, Fuel (1970), 49, 197.
2-18.   R. G. Ruberto, "Oxygen and Oxygen Functional Groups in
        Coal and Coal Liquids", presented at the EPRI Contractor's
        Conference, Palo Alto, California, May 1977.
2-19.   D. C. Cronauer and R. G. Ruberto, An Investigation of
        Mechanisms of Reactions Involving Oxygen-Containing Com-
        pounds in Coal Hydrogenation", EPRI AF-442, January 1977.
2-20.   W. Orr, "Sulfur", Handbook of Geochemistry, Vol. II/r,
        Springer-Verlag Berlin-Heidelburg, New York.
2-21.   R. Neavel, "Sulfur in Coal; It's Distribution in the Seam
        and in Mine Coals", PhD Thesis, The Pennsylvania State
        University (Geology), 1966.

2-22. H. J. Gluskoter and J. A. Simon, Illinois Geological
       Survey Circular #432.
2-23. D. C. Thorstenson and F. T. MacKenzie, Nature, 234, 543
       (1971).
2-24. P. H. Given, Fuel (1960), 31, 147.
2-25. W. Wiser, Preprints Fuel Division ACS Meeting, 20 (2), 122
       (1975).
2-26. D. D. Whitehurst, M. Farcasiu, T. O. Mitchell and J. J.
       Dickert, Jr., "The Nature and Origin of Asphaltenes in
       Processed Coals", EPRI AF-480, Project 410-1, Annual
       Report, July 1977.
2-27. D. D. Whitehurst, M. Farcasiu and T. O. Mitchell, "The
       Nature and Origin of Asphaltenes in Processed Coals",
       EPRI AF-252, Project 410-1, Annual Report, February 1976.
2-28. M. Farcasiu, T. O. Mitchell and D. D. Whitehurst,
       Chem. Tech. (1977), 7, 680.
2-29. M. Farcasiu, Fuel (1977), 56, 9.
2-30. N. C. Deno, Personal Communications.
2-31. D. D. Whitehurst, T. O. Mitchell, M. Farcasiu and J. J.
       Dickert, Jr., "The Nature and Origin of Asphaltenes in
       Processed Coals", EPRI Report, Third Annual Report Under
       Project 410-1, March 1977 - January 1979.

Chapter 3

COAL LIQUIDS

I. INTRODUCTION

The term "coal liquids" generally refers to the soluble products obtained from heating coal, usually in the presence of an H-donor solvent. This nomenclature is very misleading since the majority of these products are solid at ambient temperature, especially if the dissolution occurred without an added catalyst, as in the SRC-I process. Nevertheless, since the name is so widely used, we will also follow this convention in this book.

In this chapter, we will discuss the chemical structure of solvent-refined coals (SRC), the pyridine-soluble products with boiling points generally above $650^{\circ}F$. Knowledge of the chemical structure is valuable, from both a theoretical and practical standpoint. In theory, the structure of the soluble products provides a basis for establishing the structure of coal. Relatively few chemical or physical methods can be used to study coal itself due to its heterogeneity and insolubility, but once it is dissolved, several methods can be used to establish the chemical structure. The most relevant coal liquids for this type of study are those which have undergone a minimum amount of transformation. (As will be shown later, coal liquids obtained in some SRC-type processes meet this condition.)

From an industrial standpoint, the study of the composition of coal liquids is essential for their further processing (removing sulfur or other heteroatoms, hydrogenating rings, etc.). This must include determining the chemical nature of coal liquids and their dependence on the physical parameters of time, temperature, pressure and the type of catalyst, if any.

Because of the diverse properties of different coals, they do not respond uniformly to a particular chemical processing. Some methodology (possibly based on coal liquids) is thus essential to establish the main structural properties of coal in order to use the coal efficiently.

II.   STRUCTURAL CHARACTERIZATION OF COAL LIQUIDS

Characterization of coal liquids in the past has been primarily based on solubility characteristics. The most commonly used classification has been the pentane- and benzene-soluble materials (called oils and asphaltenes) (3-1,3-2,3-3). These terms originated from similar classifications in the petroleum industry. Recently, a pyridine-soluble benzene-insoluble fraction has also been included, often called asphaltols or pre-asphaltenes (3-4). The fraction is generally not found in petroleum feedstocks. Unfortunately, the use of extraction to separate the pentane-soluble, benzene-soluble and pyridine-soluble fractions provides very limited chemical information because the solubility of a substance is not a function of its general carbon-hydrogen skeleton and chemical functionality alone. It also depends on interactions with other substances which can act as co-solvents, such that a given compound may appear as an "oil" or an "asphaltene" depending upon the presence or absence of other substances in coal liquids. Even in very carefully performed extractions, some compounds are distributed between extract and residue.

Detailed characterization of the isolated pentane- and benzene-soluble portions of coal liquids, have often been based on fractionation by liquid chromatography using procedures developed for petroleum-derived materials followed by additional chemical methods.

Recently a method of direct fractionation of whole coal liquids was developed by the authors. This method, called sequential elution by specific solvents chromatography (SESC), utilizes liquid chromatography to isolate chemically distinct classes and was optimized specifically for coal liquids. Individual fractions thus have similar chemical functionalities but can differ in hydrocarbon skeletons depending on the type of coal used and the process conditions. An important aspect of this method is the ability to fractionate the benzene-insoluble components.

One major objective in the development of this technique was the chemical breakdown of polar substances, or asphaltols, which can constitute as much as 85% of the solvent-refined coal. Silica gel was used as the stationary phase because its mildly-acidic properties are favorable to this separation; in addition, fewer of the substances present in the mixture are irreversibly adsorbed by silica gel than by other adsorbents.

The sequence of solvents was based on the Hildebrand parameter (3-5) and the specific solubility parameters. In principle, the sequence of solvents for eluting the polar materials changes systematically, from solvents with dipole moment but no proton

acceptor properties (as in chloroform) to those that are highly
polar and have good acceptor properties (as in pyridine).

The separative technique is further optimized by alteration
of the properties of the silica gel during the fractionation. For
the first four fractions, the silica gel contains 4% water and is
highly active. After that, it is deactivated with ethanol and
maintained in this state until the end of the separation.

The result of the choice of solvents and the method of alter-
ing the activity of the silica gel is a rather unconventional
sequence of solvents. The most polar solvent (methanol) occurs in
the middle of the fractionation sequence. In the case of most
bituminous coal liquids this is a rather important advantage: the
methanol fraction is approximately the dividing line between benzene-soluble and benzene-insoluble materials, or between asphaltenes and asphaltols.

The fractions separated by this method were further studied by
spectroscopic methods, elemental analysis, molecular-weight determinations, thin-layer chromatography, thermogravimetric analysis
and other techniques.

The following sequence of solvents was used to obtain the
various fractions:

1.   Hexane
2.   Hexane/15% benzene
3.   Chloroform
4.   Chloroform/10% diethyl ether
5.   Diethyl ether/3% ethanol
6.   Methanol
7.   Chloroform/3% ethanol
8.   Tetrahydrofuran
9.   Pyridine/3% ethanol

Comparisons of results of fractionating SRC and model compounds ensured that each solvent eluted only a group of related
substances. For some model compounds, liquid chromatography was
used to observe in which solvent a given functionality class was
eluted. This helped to determine which functional groups were
represented in each fraction. But for the most part, identifica-
tion of SESC fractions was based on thin-layer chromatography (TLC)
on silica-gel plates, a more convenient method which gave quick and
reproducible results.

The use of relative retention factor R_f (the ratio of the
movement of a given compound to the movement of the solvent front)
to characterize the interaction between eluent, adsorbent and a
given substance is a common practice in TLC. For a given adsorbent, the R_f value of different compounds will depend on both the
eluent and the chemical nature of the compound.

Relative retention factors were determined for the fractions isolated from West Kentucky 9,14 SRC and it appears that the chosen solvents are indeed selective. In a given solvent, a fraction either does not move or else all its components move together with a high R_f value (0.6 or more). Each fraction has a unique corresponding solvent in which it first moves with a high R_f. With the same solvents the R_f values were determined for a number of model compounds and the same type of behavior noted. The results were used to infer the possible functionality of the SRC fractions. The major components found in the fractions are presented in Table 3-1 below. The structures of the fractions are discussed in greater detail under the section "Chemical Classes".

The chemical nature of coal liquids will be discussed in terms of the following chemical properties:

a. Elemental analyses
b. Molecular weight ranges
c. C-H skeletal structures
d. Chemical classes

A. Elemental Analyses

Coal liquids, the dissolved organic portion of the whole coal, contain mainly carbon, hydrogen, oxygen, nitrogen and sulfur. Measurements of the elemental composition provide glimpses into the characteristics of coal liquids and the severity at which they have been processed. In mild conversion of coal to soluble products, little change occurs in the carbon skeleton of the coal. For example, Illinois #6 (Monterey) coal and its SRC products have the following general formulas (calculated for 100 atoms of carbon):

Illinois #6 (Monterey)	General Formula	# Heteroatoms
		100 atoms C
Coal (75–425μ screened samples)	$C_{100}H_{88}N_{1.6}O_{13.2}S_{1.7}$	16.5
SRC (short contact time, 1-4 minutes, 800°F)	$C_{100}H_{88}N_{1.6}O_{9.1}S_{1.3}$	12.0
SRC (long contact time, 60-90 minutes, 800°F)	$C_{100}H_{89}N_{1.7}O_{5.1}S_{0.7}$	7.5

TABLE 3-1. Fractions Obtained by Mobil's SESC Procedure

Classical Description	Fraction	Major Components	General Formula for Kentucky SRC
(Oils) (1-3)	1	Saturates	
(Oils) (1-3)	2	Aromatics	$C_{24}H_{16}O, C_{24}H_{17}N$
(Asphaltenes) (3-5)	3	Polar aromatics; nonbasic N,O S-heterocyclics	
(Multifunctional Compounds) (5-9)	4	Monophenols	$C_{22}H_{16}O, C_{22}H_{17}NO$
	5	Basic nitrogen heterocyclics	$C_{30}H_{25}NO_2$
	6	Highly functional molecules (>10 wt% heteroatoms)	
	7	Polyphenols	$C_{41}H_{30}NO_3$
	8	Increasing O content and increasing basicity of nitrogen	$C_{61}H_{46}N_2O_4$
	9		
	10	Non-eluted, unidentified materials	(0-8 wt% of SRC)

Little difference exists in the number of hydrogen or nitrogen atoms per 100 atoms of carbon between the SRCs and the parent coal. After about one-hour reaction times at 800°F, nearly 60% of the sulfur and oxygen content of the parent coal is removed from the SRC. For coals of 80% C and higher the sulfur and oxygen contents are only slightly reduced in the short-contact time SRC, indicating that the chemical structures of SCT-SRCs are close to those of the parent coal.

P. Molecular weight

Liquefaction experiments at about 800°F (3-6, 3-7) indicate that coal can be fragmented into smaller pyridine-soluble molecules in a very short time. This fragmentation results from the thermal rupture of a few weak chemical bonds such as ether bonds and alkyl bridges between aromatic groups. The result is a bimodal molecular weight distribution, in which the heavy components have molecular weights greater than 2000 and light components have molecular weights between 300-900.

Figure 3-1 illustrates the change in this bimodal distribution in a Wyodak-Anderson SRC with time. After one minute, the ratio of heavy to light components is 0.65; this ratio rapidly drops to about 0.2 after 5 minutes, then to about 0.06 over the next two hours. The data for Figure 3-1 were generated using gel permeation chromatography, not SESC-type analysis, since the amount of SESC Fraction 10, the non-eluted product, is significant at low conversions.

The average molecular weight of each SESC fraction appears to depend primarily on the coal type, rather than the reaction time (or process severity). One of the most significant results of this work was finding that the average molecular weights of the individual SESC fractions did not change with the degree of conversion. In Table 3-2, average molecular weights for SESC fractions 3, 4, 5, 7 and 8 are tabulated for low and high severity SRCs from Illinois #6 (Monterey) and West Kentucky 9,14 coals. Illinois #6 (Monterey) SRC SESC-3 has an average molecular weight of 440 at low severity and 432 at high severity. For West Kentucky 9,14 SRC SESC-3, the average MW is only 315, while that of a high-severity Wyodak-Anderson SESC-3 may go as high as 660.

Field ionization mass spectrometry analysis (FIMSA) indicates that the molecular weight distribution for each SESC fraction is broad. The majority of West Kentucky 9,14 SRC SESC-3 has a molecular weight range of 200-800 (Figure 3-2). The arrows on the figures indicate weight-average molecular weights. Similarly for Wyodak-Anderson SRC SESC-3, the range is between 200-1000 (Figure 3-3). SESC fractions 4 and 5 also have broad molecular weight distributions, as indicated in Figures 3-4 and 3-5.

Although the average molecular weight of a given fraction did not appear to change, the average molecular weight of the whole SRC did decrease due to interconversion of the various fractions.

The corresponding SESC fractions of West Kentucky 9,14 and Wyodak-Anderson SRCs were found to be quite different (see Table 3-2). Wyodak-Anderson SRC has a rather uniform distribution of molecular weights for the various SESC fractions whereas West Kentucky 9,14 SRC shows a fairly monotonic increase in molecular weight with increasing functionality (SESC fraction number). An

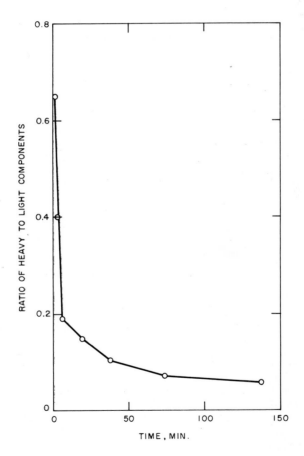

FIGURE 3-1.   Wyodak-Anderson SRC Molecular Weight Distribution
              vs. Time

TABLE 3-2. *Average Molecular Weight Determinations of SESC-SRC Fractions*

SESC Fraction	Low Severity Kentucky SRC Mobil	High Severity Kentucky SRC Wilsonville, Ala.	Low Severity Monterey SRC Mobil	High Severity Monterey SRC Mobil	High Severity Wyodak SRC HRI
3	315[a]	⎫ 309[a] and 313 ⎬	440	432	660[a]
4	300[a]	⎭	470	487	630[a]
5	470	481[a] and 476	545	492	580[a]
7	610	⎫ 715 ⎬			
8	960	⎭	1290	1142	740

[a] Chloroform was used as the solvent in the vapor pressure osmometry.

Note: Values determined by vapor pressure osmometry are number average molecular weights. All entries except those marked with an asterisk used tetrahydrofuran as the solvent in the vapor pressure osmometry.

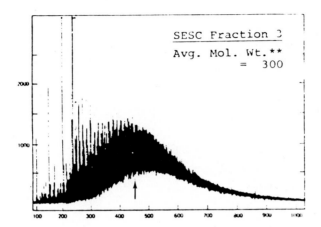

FIGURE 3-2. Field Ionization Mass Spectroscopic Analysis
of Kentucky SRC. Average molecular weight determined by vapor
phase osmometry was 300.

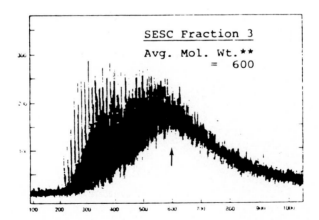

FIGURE 3-3. Field Ionization Mass Spectroscopic Analysis
of Wyodak SRC. Average molecular weight determined by vapor phase
osmometry was 600.

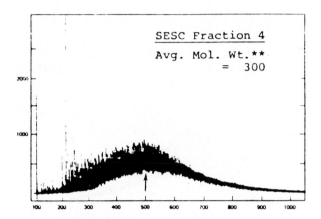

FIGURE 3-4. Field Ionization Mass Spectroscopic Analysis
of Kentucky SRC. Average molecular weight determined by vapor
phase osmometry was 300.

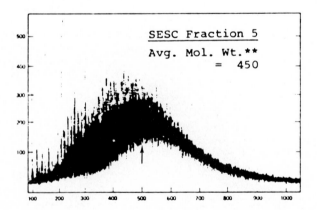

FIGURE 3-5. Field Ionization Mass Spectroscopic Analysis
of Kentucky SRC. Average molecular weight determined by vapor
phase osmometry was 450.

interesting observation was that nearly all Wyodak-Anderson SESC fractions contained more etheric oxygen than corresponding West Kentucky 9,14 SESC fractions. This allows higher oxygen content for the same SESC fraction.

The very high molecular weight compounds, (2000), are believed to be concentrated in SESC fractions 9 and 10. As much as 70-85% of the initial product of bituminous coals are benzene-insoluble/ pyridine-soluble. These products are highly functional.

C.   C-H Skeletal Structures

1.   Aromaticity of SRCs

In determining the gross skeletal structure of coal liquids or fractions thereof, establishing the aromatic carbon and hydrogen contents is essential.

With the advent of ^{13}C-NMR, the aromatic content of SRCs can now be determined directly. Solid state CP-^{13}C-NMR is also useful for analysis of coals directly. This latter technique, however, does not provide the high resolution found for solution spectra.

High resolution has been very informative in structural assignments for petroleum-derived fuels and many correlations of chemical shift and structure type have been developed. Such correlations may be valid for coal distillates but for high-boiling (high molecular weight) coal fractions, such as SRC, they should be used with caution.

Over the past 15 years many papers used narrowly-defined chemical shift regions in ^{1}H-NMR as indicators of chemical structure based on reports by J. K. Brown and W. R. Ladner (3-8,3-9). The region of the spectra between 0.5 and 4 ppm is assigned to specific types of aliphatic hydrogens (6-8 ppm indicate aromatic ring protons). These assignments were derived from materials of petroleum origin and are reasonably valid for such materials. A recent publication (3-10) proposed the same kind of narrow chemical shift definition for ^{13}C-NMR spectra. We believe that use of such correlations is not advisable for high molecular weight coal liquids and that in some cases it can be misleading because of the high content of fused aliphatic rings in coal liquids.

Adsorptions in the region of 130-150 ppm have been correlated to structures such as the following:

OH
↙ 155ppm

158ppm
↙ O

Chemical shifts >150 ppm have been assigned as aromatic carbon
containing O, N, or S substituents.   As will be shown later,
structures of polycondensed saturated hydrocarbons bonded to
aromatic structures are highly probable in some coal liquids.   If
one considers such structures, the ^{13}C-NMR results become less
definitive.   For example, compounds such as 1-phenyladamantane and
benzotetrahydrodicyclopentadiene have ^{13}C-NMR and proton NMR
spectra which fall outside correlations developed for petroleum
derived materials.

^{13}C-NMR CHEMICAL SHIFTS

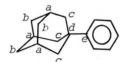

1-PHENYL ADAMANTANE

(a)	29.0 ppm
(b)	36.9 ppm
(c)	43.3 ppm
(d)	36.2 ppm
(e)	151.2 ppm

One important observation in the above structure is that a
chemical shift as large as 151 ppm can be correlated to a
C_{ar}-C_{aliph} bond and does not have to be C_{ar}-O (or other
heteroatom, such as N) bond.
In ^{1}H-NMR, adsorption between 2 and 3 ppm is usually
considered to be the "benzylic" region.   The 1-phenyladamantane
(see Figure 3-6) obviously does not contain any benzylic hydrogen
and yet there is significant adsorption above 2 ppm.
As shown in Figure 3-7, benzotetrahydrodicyclopentadiene has
38% of the protons in the region 2-3.3 ppm.   The true benzylic
protons would account for only 21% of the protons.
These observations support the conclusion that the use of
chemical shifts in both ^{13}C-NMR and ^{1}H-NMR to determine specific
functionalities in coal liquids should be attempted with caution.
The main use of these methods is, we believe, the determination of
the percentages of aromatic carbons and hydrogens, and in the
following text ^{1}H-NMR and ^{13}C-NMR were used only for that purpose.

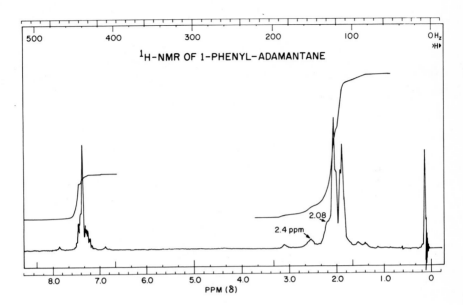

FIGURE 3-6. ^{1}H-NMR of 1-Phenyladamantane

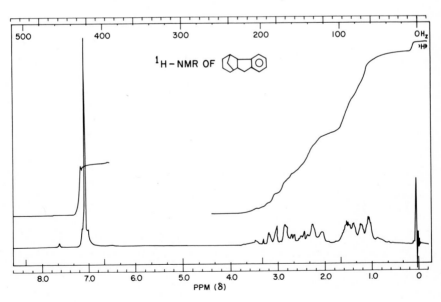

FIGURE 3-7. ^{1}H-NMR of Benzotetrahydrodicyclopentadiene

2. Degree of Aromatic Ring Condensation

Although ^{13}C-NMR provides information on the total aromatic
carbon content, the essential continuum of chemical shifts in that
region of SRC spectra precludes the distinction between condensed
and non-condensed aromatic carbons. One technique which shows
promise in providing quantitative information of this type is
non-aqueous polarographic reduction (3-11). Benzene nuclei are
inaccessible to direct polarographic reduction as their estimated
E_{1_2} is about - 3.5 V, far outside the range of any electrolyte-
solvent system presently available. Aromatic systems which are
more complex than benzene show polarographic reduction potentials
at voltages which depend upon the complexity of the conjugation.
In general, the reduction potential decreases as the number of
double bonds in conjugation increases. This trend is illustrated
below; half-wave potentials relative to a standard calomel
electrode are given below the compounds.

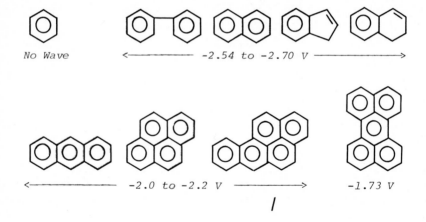

Unconjugated aromatic systems show half-wave potentials (E_{1_2})
for the simplest, complete aromatic system

Like benzene, phenothiazine gives no wave.

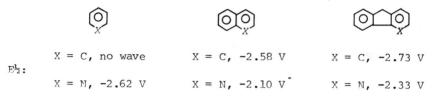

-2.6 V (compared to naphthalene's $E_{\frac{1}{2}}$ of -2.58 V)

Replacing an aromatic carbon by a nitrogen atom lowers the half-wave potential by about 0.5 V.

$E_{\frac{1}{2}}$:

X = C, no wave   X = C, -2.58 V   X = C, -2.73 V

X = N, -2.62 V   X = N, -2.10 V   X = N, -2.33 V

For quantitative polarography, the Ilkovic equation and relationships derived therefrom are of fundamental importance. The general form of this equation is

$$i_d = knD^{1/2}C_o m^{2/3} t^{1/6}$$

where i_d = diffusion current in amps

n = number of electrons/molecule taking part in the process
D = diffusion coefficient in cm^2/sec
C_o = concentration of active substance in mmole/l
m = mercury flow rate in mg/sec
t = mercury drop time in sec

When m and t are held constant, then i_d is proportional to C_o and this relationship has become the basis for quantitative work, especially when internal standards are used.

For example, the reduction of naphthalene occurs at -2.5 V. The reaction corresponds to the addition of 2 electrons via a 1,4 addition mechanism forming 1,4-dihydronaphthalene which is not further reduced.

$$\xrightarrow[+2H^+]{+2e^-}$$

$$\frac{i_d}{C_o} = 19.0$$

The diffusion constant calculated from the current and concentration (i.e., i_d/C_o = 12.2) will be used to calculate the concentration of "naphthalene-like species" in the coal liquids at this $E_{\frac{1}{2}}$.

Theoretically, the diffusion current is also related to the molecular weight by the following expression

$$i_d = K \sqrt{\frac{1}{(3M)^{1/3}}}$$

The sensitivity of i_d to molecular weight is, therefore, only a minor consideration.

At the present state of the development of this technique, no distinction can be made between naphthalene and phenanthrene ring structures. Biphenyls would also reduce at the same voltage. Therefore, in the following text the degree of ring condensation is reported in terms of fluoranthene, pyrene, anthracene, and naphthalene and/or phenanthrene contents.

The content of condensed rings larger than phenanthrene is relatively small, ranging frcm 0.0 to 5.9% of the carbon found in "pyrene-like" rings. Thus, the large majority of the carbon is found in ring systems less condensed than pyrene even in severely processed SRCs.

3. C-H Skeleton - Average Structures

The major elements in coal and coal liquids are carbon and hydrogen. The number of heteroatoms (N,O,S) per 100 atoms of carbon is no greater than 15 for a bituminous coal (in Elemental Analysis) and decreases to about 7-8 in long-contact-time SRC. At the same time, the atomic ratio of carbon to hydrogen ranges from 0.7-0.9 in the majority of coal liquids. This suggests that changes in the C-H skeleton during liquefaction are at least as important as the changes in chemical functionality.

Major skeletal differences are noted if one compares the fractions of common functionality, e.g., SESC-3 (ethers and compounds with nonbasic nitrogen) from different coals. Table 3-3 shows the broad variation in C-H structure for SRCs from the same coal but under different conditions, and for SRCs from different coals. By using the various compositional analyses available today, an "average" chemical structure for a coal liquid or a specific fraction can be derived, giving a general picture of the types of skeletal structures present. The weight average molecular weight, the elemental analysis, and the percent aromatic carbon and hydrogen are needed, as well as polarographic measurements of the degree of aromatic ring condensation, infra-red spectra, etc. Recent studies on selective oxidation of coals and SRCs (3-13) have shown that acyclic aliphatic structures are relatively important. Methyl substituted aromatics account for less than one methyl for each 3 "average" molecules. Thus, methyl substitutions need not be considered. Within these boundaries, a structure can be drawn which represents the "average" structure of molecules found in such a material.

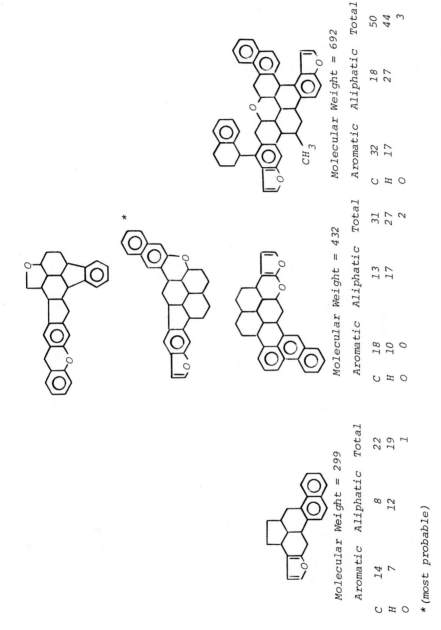

	Molecular Weight = 299		
	Aromatic	Aliphatic	Total
C	14	8	22
H	7	12	19
O	0	1	1

	Molecular Weight = 432		
	Aromatic	Aliphatic	Total
C	18	13	31
H	10	17	27
O	0	2	2

	Molecular Weight = 692		
	Aromatic	Aliphatic	Total
C	32	18	50
H	17	27	44
O		3	3

* (most probable)

Figure 3-8. Structures for High and Low Molecular Weight Illinois #6 (Monterey) SESC-3

TABLE 3-3 . *Fraction SESC-3 From Various Sources*

Fraction SESC-3	% in SRC	No. Avg. MW	% Aromatic H	% Aromatic C	General Formula
Kentucky SRC Wilsonville, Ala. (long contact time)	21.5	300	83	81	$C_{24}H_{16}O$ $C_{24}H_{17}N$
Kentucky SRC (mild conditions short contact time)	6.6	300	28	≤ 60	$C_{21}H_{21}O$
Monterey SRC (long contact time)	18.0	440	39	58	$C_{31}H_{27}O_{1-2}$
Wyodak SRC (HRI)	20.8	600	48	60	$C_{49}H_{42}O$

In view of the wide variation in the molecular weights for coal
liquids, one might suspect that the structures one would draw for
low and high molecular weight compounds could be grossly different.
In fact, they are not, as shown in Figure 3-8 for Illinois #6 (Monterey) SESC-3. In this figure, the structures were developed assuming a 300-molecular-weight (the low end of the molecular-weight
profile), an "average" molecular weight of 430, and a 700-molecular
weight (the high end of the molecular weight profile). These structures were derived using the same analytical data other than molecular weight. It can be seen that the kind of structures which must
be considered are very similar. The corresponding formula average
structures for two SESC-3 fractions of West Kentucky 9,14 SRCs, produced at short and long times, are shown in Figure 3-9.

The following example shows how such structures are derived.
Consider a fraction with an average molecular formula of $C_{31}H_{27}O_2$
and the aromatic moiety is $C_{18}H_{10}$ (as determined by ^{13}C-NMR and
^{1}H-NMR), the following aromatic structures are equally probable:

VARIANT 1

VARIANT 2

VARIANT 3

etc., etc.

Such structural studies indicate monoaromatic, benzofuran and naphthalene are the principal types of aromatic rings in short-contact-time liquefaction products from West Kentucky 9,14 and Wyodak-Anderson coals.

In principle, the derivation of "average" structures can pro-vide a great deal of insight into the intrinsic limitations of conversion of coals to conventional fuels. For example, the struc-ture shown in Figure 3-9 indicates that the dehydrogenation of aliphatic rings to produce polyaromatics is quite probable. Also the conversion of such structures to conventional paraffinic fuels (e.g., diesel) will be very difficult because of the extensive ring openings that will be required. Gasoline, which is highly aromatic, could be easier to generate.

Average Structures

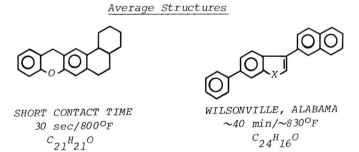

SHORT CONTACT TIME
30 sec/800°F
$C_{21}H_{21}O$

WILSONVILLE, ALABAMA
~40 min/~830°F
$C_{24}H_{16}O$

FIGURE 3-9.   SESC-3 Kentucky SRC

D. Chemical Classes - Analyses and Proposed Structures of SESC
Fractions

SRC coal liquids are complicated mixtures of many chemical
substances and contain molecules of diverse chemical
functionalities. The primary functionalities associated with
oxygen, sulfur and nitrogen are as follows:

a)   oxygen in phenols, acyclic or heterocyclic ethers, and as
carboxylic acids.
b)   sulfur in aromatic heterocyclic compounds and possibly
thiophenols.
c)   nitrogen in pyrroles or pyridine-type heterocycles

Only a few single components can be separated from SRCs and
their amounts are usually insignificant as a fraction of the
total. Separating the SRC into fractions of common functionalities
has been found to be a more meaningful method of developing
structural information about SRC liquids. The use of sequential
elution by specific solvents chromatography (SESC), as described
earlier in this chapter, allows one to partition the SRC into ten
fractions, each containing common major chemical functionalities
as was shown in Table 3-1. Detailed analyses of the various
fractions of West Kentucky 9,14, Illinois #6 (Monterey) and
Wyodak-Anderson SRCs are given in Tables 3-4 and 3-5.
Some of these SESC fractions have been studied in enough
detail that molecular structures can be proposed as described in
the following text.   The experimental techniques used in these
studies are as follows:

- vapor phase osmometry for molecular weight determinations
- elemental analysis
- ^{1}H-NMR for aromatic hydrogen content
- ^{13}C-NMR for aromatic carbon content
- gas chromatography - for light components
- thin-layer chromatography and infra-red absorption (i.r.)
  - for functionality
- polarography for an indication of aromatic ring
  condensation

In general, the functionality of the various fractions increases
with increasing fraction number; the opposite trend is seen for
aromatic content, as shown in Figure 3-10.

TABLE 3-4. Experimental Data on Monterey SRC

Sample	Elemental Analysis					Mol Wt By VPO in THF	% Ar H	% Ar C	General Formula	Rx Conditions			Yield % MAF Coal
	C	H	O	N	S					Time	Temp	Solv	
AC-1-SRC	80.50	5.92	8.90	1.99	2.73	-	38	-	$C_{100}H_{88}N_{1.5}O_{8.3}S_{1.3}$		4 min/800°F/TMN[a]		77.9
AC-2-SRC	85.19	5.95	3.73	1.63	1.56	-	47	-	$C_{100}H_{84}N_{1.6}O_{3.3}S_{0.7}$		90 min/800°F/TMN		64.0
AC-3-SRC	79.13	5.83	9.59	1.50	2.75	-	40	-	$C_{100}H_{88}N_{1.6}O_{9.1}S_{1.3}$		4 min/800°F/ SS[a]		82.8
AC-4-SRC	85.0	5.91	5.82	1.67	1.51	-	47	-	$C_{100}H_{84}N_{1.7}O_{5.1}S_{0.7}$		90 min/800°F/ SS		68.6
Fraction 3													
AC-2-3	86.35	6.32	4.64	0.95	1.81	432	39	58	$C_{31}H_{27}O_{1.25}N_{0.3}S_{0.2}$				8.8
AC-3-3	82.1	6.91	10.96			440	-	-	$C_{30}H_{30}O_3$				5.3
Fraction 4													
AC-2-4	85.1	6.08	6.32	1.02	1.46	487	-	-	$C_{34}H_{29}O_{1.9}N_{0.3}S_{0.2}$				14.1
AC-3-4	81.2	6.6	8.3	1.0	2.4	470	43	59	$C_{32}H_{31}O_{2.4}N_{0.3}S_{0.4}$				20.2
Fraction 5													
AC-2-5	80.5	6.07	9.43	-	2.51	492			$C_{33}H_{30}O_{2.9}S_{0.4}$				6.5
AC-3-5	77.11	6.25	12.99			545			$C_{35}H_{39}O_{4.2}$				14.0
Fraction 8													
AC-2-8	79.78	5.32	11.72			1142			$C_{76}H_{60}O_{8.4}$				6.9
AC-3-8	78.6	5.52	11.14		2.34	1290			$C_{84}H_{71}O_9$				6.6

[a] TMN is (53%) tetralin (47%) methylnaphthalene and SS is our standard synthetic solvent.

TABLE 3-5. Composition of Kentucky and Wyodak SRC's

Sample	% in SRC	% C	%H	%N	%S	% O	M/W Solvent	% Ar H	% Ar C	General Formula
Wilsonville Ky. SRC[a]		87.6	4.8	2.0	0.8	3.4	-	70	-	$C_1H_{0.65}N_{0.02}O_{0.03}$ [d,e,f]
AC-5 SRC[b]		84.08	5.62	1.78	1.51	6.68	-	≈37	<57	$C_1H_{0.8}N_{0.02}O_{0.06}$ [d,e,f]
Wyodak SRC[c]		86.81	5.76	1.30	0.34	5.46	-	52	62	$C_1H_{0.8}N_{0.02}O_{0.05}$ [d,e,f]
Wyodak SRC AC-6 (short contact)		80.18	6.53	0.85	0.55	11.75	-	30	-	
Wyodak SRC AC-7 (long contact)		86.16	6.52	1.35	0.37	5.08	-	44	-	
Fraction 3										
Wilsonville SRC	21.5	91.8	5.38	0.84	0.8	1.06	300/CHCl₃	83	81	$C_{24}H_{16}O, C_{24}H_{17}N$
AC-5 SRC	6.6	84.43	6.90	0.68	1.35	5.38	300/CHCl₃	28	-	$C_{21}H_{21}O$
Wyodak SRC	20.8	89.68	6.15	1.08	-	∼3	660/CHCl₃	48	-	$C_{49}H_{42}O, C_{49}H_{49}NO$
Fraction 4										
Wilsonville SRC	13.6	86.3	5.33	2.32	0.8	5.2	300/CHCl₃	63	74	$C_{22}H_{16}O, C_{22}H_{17}NO$
AC-5 SRC	7.0	79.08	6.68	0.74	1.12	11.5	300/CHCl₃	28	-	$C_{20}H_{20}O_2$
Wyodak SRC	20.0	85.2	6.07	1.97	-	6.3	630/CHCl₃	47	-	$C_{45}H_{38}NO_2$
Fraction 5										
Wilsonville SRC	13.5	80.6	5.64	3.81	1.31	8.3	∼450/CHCl₃	49	64	$C_{30}H_{25}NO_2$
AC-5 SRC	17.5	79.62	6.32	1.98	1.06	10.65	470/CHCl₃	-	60	$C_{31}H_{30}O_3, C_{31}H_{31}NO_3$
Wyodak SRC	10.7	79.54	6.57	1.82	0.76	10.99	500/THF	47	-	$C_{40}H_{38}NO_4$

Fraction 6										
Wilsonville SRC	4.7	73.32	5.18	5.2	5.92	10.4	–	–	50	
AC-5 SRC	6.3	75.7	6.5	2.7	3.0	12	–	–	–	
Wyodak SRC	4.4	74.26	6.01	2.6	0.65	11.7	–	36	–	for MW ~600 $C_{41}H_{30}NO_3$
Fraction 7										
Wilsonville SRC	2.13	82.43	5.01	2.53	0.93	7.67	–	–	–	$C_{41}H_{35}NO_4$
AC-5 SRC	11.0	80.9	5.75	1.88	1.75	10.0	610/THF	45	–	
Wyodak SRC	4.7	84.31	5.75	1.94	–	6.12	–	–	–	for MW ~600 $C_{42}H_{33}NO_2$
Fraction 8										
Wilsonville SRC	13.0	81.9	5.14	2.74	1.7	7.7	–	60	73	for MW ~900 $C_{61}H_{46}N_2O_4$ $C_{61}H_{46}NO_5$
AC-5 SRC	26.3	78.4	7.05	1.54	1.36	11.5	960/THF	28	–	$C_{63}H_{67}NO_7$
Wyodak SRC	16.3	81.6	5.84	1.58	–	9.95	740/THF	–	–	$C_{50}H_{43}NO_5$

[a] Kentucky SRC Wilsonville, Alabama.

[b] Kentucky SRC (our run #10, 30s/800°F, synthetic solvent).

[c] Wyodak SRC, HRI run 177-114-2B.

[d] See text for the description of the run.

[e] Rel. no. of atoms for 1C.

[f] Rel. no. of atoms for 1C in Kentucky coal $C_1H_{0.82}N_{0.017}O_{0.094}$; in Wyodak $C_1H_{0.86}N_{0.01}O_{0.18}$

SESC FRACTIONS 1-9

Fraction 1 (eluent hexane) - This fraction constitutes a small
portion of a West Kentucky 9,14 SRC (about 0.4%), after long reaction times; it contains a mixture of saturated acyclic and cyclic
hydrocarbons.

Fraction 2 (eluent hexane/15% benzene) - Aromatic and partially
hydrogenated aromatic hydrocarbons have been identified in this
fraction, also for West Kentucky 9,14 coal, which contains primarily molecules with 18 or more carbon atoms.
      The aromatic proton content was found to range from 30-85% as
a function of the severity of the liquefaction process; this
phenomenon is due to the aromatization of hydrocarbons as the extent of liquefaction increases, thus increasing the aromatic proton
content. For the most part, fraction 2 is not a major constituent
at short times. Gas chromatography has verified that none of the
hydrocarbons present in fraction 2 appear in any subsequent fractions.

Fraction 3 (eluent chloroform) - This fraction is one of the major
fractions in many long contact time SRCs, - over 20% of West Kentucky 9,14 and Wyodak-Anderson-derived coal liquids, and 14% of
those from Illinois #6 (Monterey).
      Possible average structures for West Kentucky 9,14 coal-de-
rived fraction 3 and supporting experimental data are presented in
Figure 3-11 (a)-(c). The aromaticity of this fraction, as in fraction 2, is a function of the severity of the liquefaction process.
      Based on thin-layer chromatography of model compounds, the
components of this fraction probably contain no more than one
heteroatom per molecule. The average molecular weight of this
fraction of West Kentucky 9,14 is 300. These data lead to the general formulas proposed in Figure 3-11(a). Note that half the
molecules contain no heteroatoms.
      The relative retention factors (R_f) for model compounds indi-
cate that structures I and II in Figure 3-11(b) are possible for
the portions of fraction 3 which contain heteroatoms. Such structures have extreme sensitivity toward oxidation. This is in agreement with their chemistry as they oxidize rather easily even at
room temperature (3-12). An additional structure, such as a partially hydrogenated aromatic may also contribute to this sensitivity.
      Based on the above discussion, the type of structure shown in
Figure 3-11(c) is a reasonable representation of the heteroatom-
containing constituent in this fraction. Some molecules may also
contain hydroaromatic substituents, and occasionally, substituted
phenanthrene or anthracene types could appear. The double bond in
the five-membered ring is necessary to explain the chemistry of
this fraction.

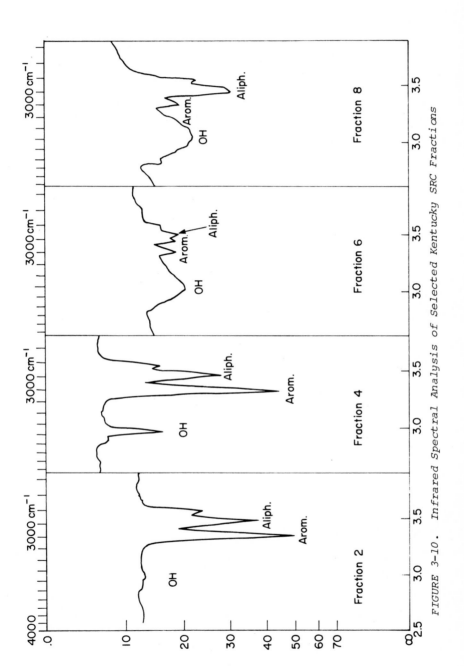

FIGURE 3-10. Infrared Spectral Analysis of Selected Kentucky SRC Fractions

C 91.8%   H 5.4%   S 0.8%   N 0.85%   O 1.1%
MW = 300   (CHCl₃ or THF)

25% Oxygen compounds $C_{24}H_{16}O$
20% Nitrogen compounds $C_{24}H_{17}N$
10% Sulphur compounds $C_{24}H_{16}S$
45% Hydrocarbons $C_{22-24}H_{16-18}$

(a)

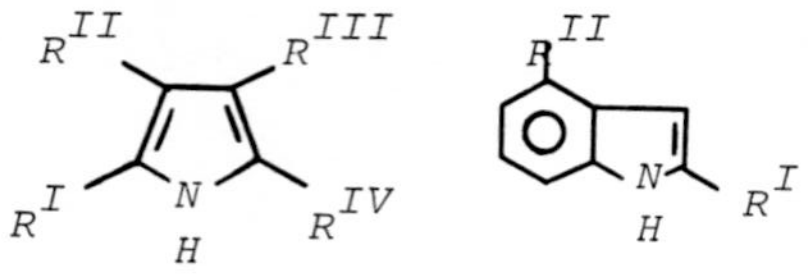

for N	for O
For Kentucky SRC in Frac. #3	83% Aromatic H
	81% Aromatic C

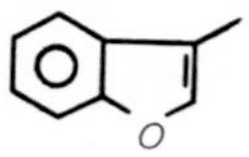

IR: no OH, very strong aromatic absorption
Polarography: reduction in the region -1.8 to -2.6V
TLC: R_f as for compounds of types shown.

(b)

Possible average structure for heteroatom containing portion of fraction 3 in Kentucky SRC - $C_{24}H_{16}O$

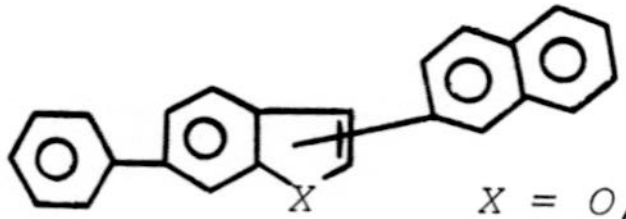

X = O, S, NH

Wilsonville, Alabama ~40 min/<830°F   (c)

Composition $C_{100}H_{82}N_{1.7}O_{9.4}$

No. polycondensed	Coal	4
saturated rings/100C	SESC-3	
	SCT	2

Structure
SESC-3
SCT-SRC

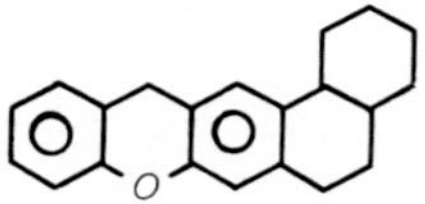

(d)

FIGURE 3-11. Kentucky SRC - Fraction 3

The same approach, used above for West Kentucky 9,14 coal liquid fraction 3 can be used to obtain possible average structures of the heteroatom-containing species in fraction 3 in short-contact time (Figure 3-11(d)) for Illinois #6 (Monterey) and Wyodak-Anderson coals.

Fraction 3 of Illinois #6 (Monterey) SRC, constitutes about 14% of the total SRC yield (% MAF coal) after 90 minutes and only 6% after 4 minutes. Based on the general formula ($C_{31}H_{27}O_2$) derived from elemental analysis and molecular weight determination by VPO (Table 3-5), and on the aromatic hydrogen and carbon contents, three aromatic moiety structures were considered (Figure 3-12). The aliphatic portions were then filled in to arrive at three possible average structures for Illinois #6 (Monterey) SESC-3 at long-contact time, two of which are unlikely.

$$General\ Formula\ C_{31}H_{27}O_2$$

$$Aromatic\ Moiety\ C_{18}H_{10-11}(^1H\text{-}NMR,\ ^{13}C\text{-}NMR)$$

Aromatic Structures Considered

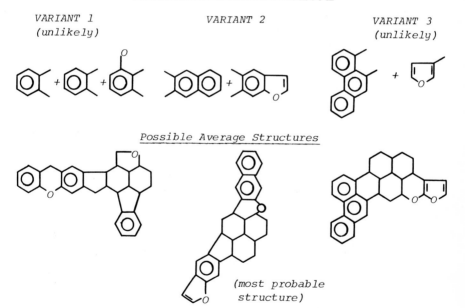

Possible Average Structures

FIGURE 3-12. Possible average structure for a long contact time fraction SESC-3 of Monterey coal.

The general formula for Wyodak-Anderson SESC fraction 3 is $C_{43}H_{88}O_2$, using weight average molecular weight. This is slightly different from the formula in Table 3-4 which used number-average molecular units. In arriving at possible average structures for Wyodak-Anderson SESC-3, one must realize the lack of aliphatic hydrogens in the "example molecule", which leads to more aliphatic carbon-carbon bonding and to carbons bonded to only one hydrogen. This is only possible in three-dimensional polycondensed saturated structures as shown in Figure 3-13.

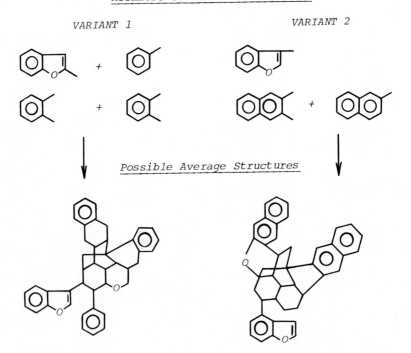

General Formula $C_{43}H_{38}O_2$

Aromatic Moiety $C_{26}H_{18}$ (^{1}H-NMR, ^{13}C-NMR)

Aromatic Structures Considered

VARIANT 1   VARIANT 2

Possible Average Structures

FIGURE 3-13. Possible Average Structures for Wyodak SESC-3.

In choosing the most probable isomeric structure for the aliphatic portion, thermodynamic stability is an important consideration. However, the most thermally stable isomeric structure for the aliphatic portion is not necessarily the only one present, although very strained (high energy content) structures are not likely to have survived the liquefaction reaction. For an indication of possible (sufficiently stable) structures of the aliphatic portion, empirical force field (strain) calculations can be used (3-13, 3-14).

For Wyodak-Anderson SESC-3, if the substituents (the aromatics and oxygen) are replaced by H-atoms, the corresponding aliphatic hydrocarbons will have the molecular formula $C_{17}H_{28}$ (i.e. C_nH_{2n-6}) as in tetracycloalkane. Force field calculations for tetracycloalkane have not been performed but have been for some tricycloalkanes ($C_{11}H_{18}$). Based on the experimental data and the above discussion, structures of the type shown in Figure 3-11(c) with some hydroaromatic constituents in a small number of molecules are likely heteroatom-containing substituents of this fraction. Of course, the position of the substituents can vary and occasionally, substituted phenanthrene or anthracene types may appear. The double bond in the five-membered rings, which is more reactive than in normal aromatic compounds, is also indicated in this structure.

Figure 3-14 shows a comparison of the "average" structures derived for a variety of coals. It can be seen that the bituminous coals have more planar condensed aliphatic rings but the subbituminous coal is significantly different and contains some three-dimensional condensed ring structures.

As will be discussed later, the aromaticity of this fraction is also a function of the severity of the liquefaction process.

A very intriguing question is the structure of the hydrocarbons present in this fraction with approximate chemical formulas of $C_{22-24}H_{16-18}$ (Figure 3-11 (a)). Gas chromatographic analysis of this fraction, even at temperatures up to 840°F, gave no component eluted from the column, a rather unexpected result since the molecular weight of this fraction is around 300. This indicates that these hydrocarbons are different from those in Fractions 1 and 2; they must therefore be polar or possess other peculiarities of structure to explain their behavior. Four classes of hydrocarbons appear possible: fulvenes, azulenes, substituted derivatives of p-quinodimethane and cyclophanes. Further work is needed to precisely identify the structures.

COAL
RANK
COMPOSITION

	Kentucky Bituminous $C_{100}H_{82}N_{1.7}O_{9.4}$	Hiawatha Bituminous $C_{100}H_{84}N_2O_{9.5}$	Monterey Bituminous $C_{100}H_{88}N_{1.6}O_{13.2}S_{1.74}$	Wyodak Subbituminous $C_{100}H_{86}NO_{18}$
# Polycondensed Coal ⎤ SESC-3 ⎟ 100 Carbons ⎦ SCT[a]	4 2	6 4	6 4	10 5

Polycondensed ⎤ Coal
Saturated Rings/ ⎟ SESC-3
100 Carbons ⎦ SCT[a]

Estimated
Average
Structure
SESC-3
SCT-SRC

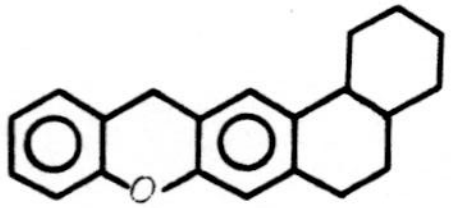

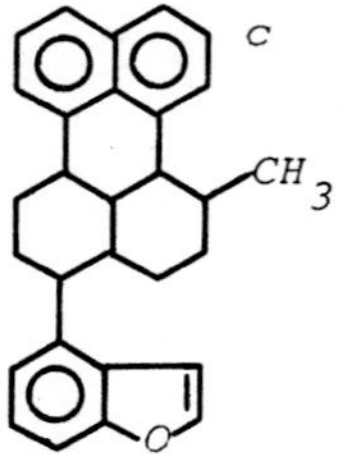

| g H Consumed/ 100g Coal During SCT ⎤ Liquefaction ⎟ (Normalized to LCT[b] ⎦ 100% Conversion) | 0.43 0.96 | | 0.63 2.03 | 0.89 2.80 |

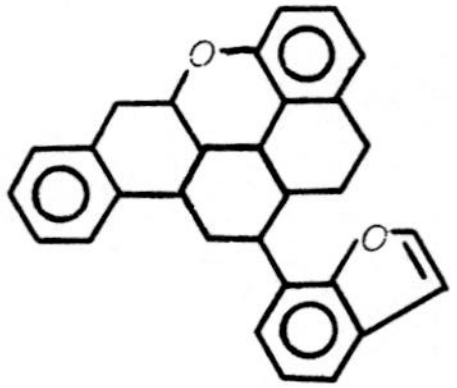

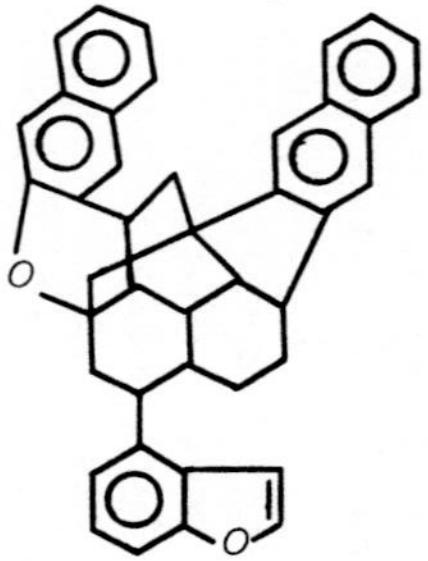

g H Consumed/
100g Coal During SCT ⎤
Liquefaction ⎟
(Normalized to LCT[b] ⎦
100% Conversion)

[a] SCT = Short Contact Time <2 minutes.
[b] LCT = Long Contact Time >30 minutes.
[c] Ca. 30 minutes contact time.

FIGURE 3-14. Comparison of Structures and Hydrogen Consumption of Four Coals

Fraction 4 (eluent chloroform/10% diethyl ether) - This fraction
is the predominant end product of subbituminous coals which contain high phenol contents (3-21). The skeletal structures are
similar to those of fraction 3. For a West Kentucky 9,14 SRC,
this fraction has an average formula of $C_{24}H_{17.8}O_{1.08}S_{0.08}$. Based
on TLC and LC comparisons with model compounds, each molecule contains one oxygen as phenolic OH and about half contain nitrogen in
indole form. The phenolic groups in long-contact-time SRCs are
believed to be attached to non-condensed aromatic rings.

Fraction 5 - (eluent diethyl ether/3% ethanol) - This fraction in
the SRC is rich in nitrogen (4% for West Kentucky 9,14 SRC), and
based on TLC studies the nitrogen appears to be basic only, as in
pyridine derivatives. Diphenols can also be eluted in this fraction. The oxygen present in molecules containing basic nitrogen
must be in etheric form.
      Experimental results and possible average structures for this
fraction in different West Kentucky 9,14 SRCs are presented in
Figure 3-15.
      All the fractions eluted up to this point are soluble in hot
benzene and represent what is usually named asphaltenes. If the
solvent-refined coal is extracted with cold benzene, fraction 5 is
less soluble. In the case of some SRCs (with high oxygen content),
this fraction is totally insoluble in benzene.

Fraction 6 - (eluent methanol) - This fraction is rather small,
but has an unusually high content of heteroatoms (over 17% by
weight). From TLC studies and considering that none of the other
fractions is eluted by methanol, this fraction probably contains
highly polar substances. Materials such as phenolphthalein eluted
in this portion as judged by model compound studies.

Fraction 7 - (eluent chloroform/3% ethanol) - The average molecular formula of this fraction has a high oxygen content (Table 3-1)
and average molecular formula $C_{42}H_{31}NO_3$ for the West Kentucky 9,14
SRC. TLC and i.r. studies indicate that oxygen is primarily present as phenolic OH.

Fraction 8 - (eluent tetrahydrofuran/3% ethanol) - The average
molecular formula of this fraction for the SRC discussed here is
$C_{61}H_{46}N_2O_4$ with the majority of oxygen present as phenolic OH
(i.r. spectroscopy). In West Kentucky 9,14 SRC, the aromatic H is
60%; aromatic C is 73%. An average structure is shown in Figure
3-16.

Run Conditions	*30 sec/808°F*	*40 min/860°F*
Average Formula	$C_{31-32}H_{29-30}NO_2$	$C_{30-31}H_{26}NO_2$
Aromatic Moiety	$C_{19-20}H_{10-11}\left(\begin{smallmatrix}{}^{1}H\text{-}NMR\\ {}^{13}C\text{-}NMR\end{smallmatrix}\right)$	$C_{19-20}H_{12-13}$
Aliphatic and Benzylic C	C_{11-12}	C_{11-12}
Benzylic (or other 2-3 ppm)	H_{10}	H_{6-7}
Aliphatic	H_{10}	H_{7-8}

N as in 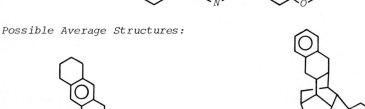 *(basic) TLC, model compounds*
O ether or as in furan

Polarography: polycondensed aromatic, traces only

Aromatic substituents considered:

Possible Average Structures:

$C_{31}H_{31}NO_2$ $C_{30}H_{26}NO_2$

Aromatic	$C_{19}H_{11}$	*Aromatic*	$C_{19}H_{12}$
Benzylic	H_9 C_{11}	*Benzylic*	H_6 C_{11}
Aliphatic	H_{10}	*Aliphatic*	H_8

Figure 3-15. West Kentucky 9,14 SRC - SESC Fraction 5

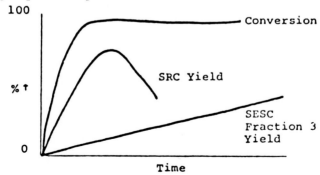

$$C_{61}H_{46}N_2O_4$$ 73% Aromatic C

900 molecular weight   60% Aromatic H

*FIGURE 3-16. Average Structure for West Kentucky 9,14
Asphaltol (SESC-8).*

Fraction 9 - (eluent pyridine/3% ethanol) - This fraction has an
elemental analysis similar to fraction 8 but presumably a higher
molecular weight.

In fractions 7,8 and 9, the quantity of polycondensed aromatic
structures (larger than naphthalene) is rather small based on
polarographic studies and on the percentage of aromatic C and H
(see data for fraction 8).

The major difference between fractions 7 - 9 is the number of
phenolic -OH groups per average molecule (which increase in the
heavier fractions).

III. CHANGES IN SESC FRACTION DISTRIBUTION WITH DEGREE OF
CONVERSION

As previously discussed (3-6), there is always a problem in
the choice of a parameter to be used for assessing the extent of
reaction in coal liquefaction. This can be seen from the
following stylized figure:

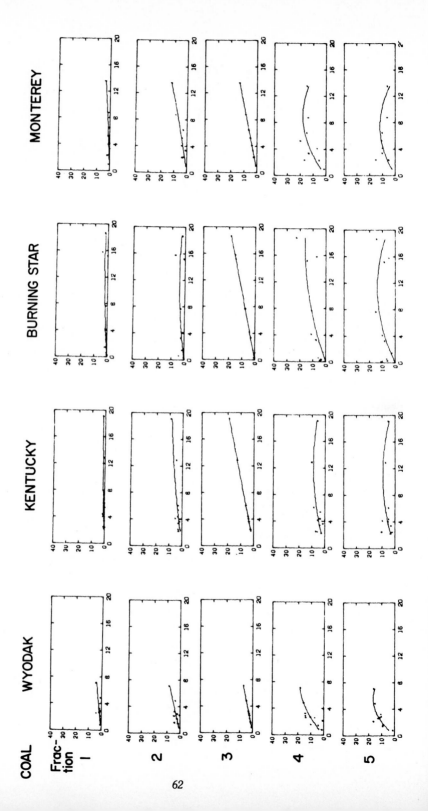

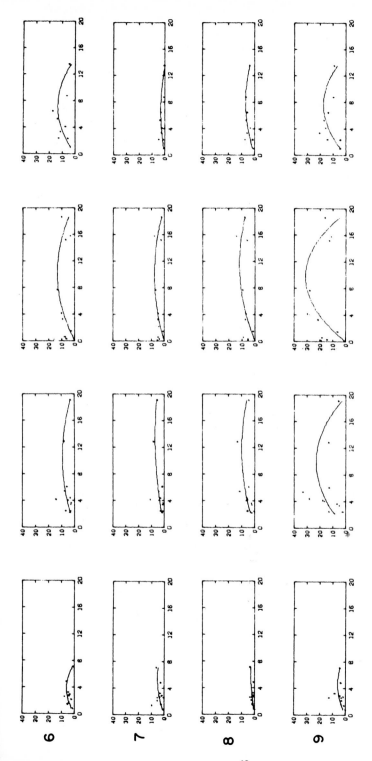

FIGURE 3-17. Fraction Yield vs. Fraction 3 Yield at Low Temperature (800°F) (cont.)

Time is not a good parameter because there are both very fast and very slow processes; conversion is also poor because it increases rapidly and then is constant; SRC yield passes through a maximum so even in a given run a certain SRC yield will not correspond to a unique time or conversion. It has been found that the yield of fraction 3 increases smoothly and steadily in conversions of bituminous coals and this value can be used as an index of extent of reaction. For lower rank coals, fraction 4 (monophenols) is more meaningful.

There are also two ways of presenting the amounts of SESC fractions obtained. Usually fraction distributions are normalized to 100%, i.e., fractions as percent of total SRC. This is useful in many cases but can be misleading when the total SRC passes through a maximum. A more appropriate presentation is one in which fractions are expressed as percent of MAF feed coal, i.e., yields of fractions based on MAF coal.

In Figure 3-17 the yields of the various fractions are plotted against the yield of fraction 3.

Within each figure the scales of all plots are the same, to facilitate visual comparison.

Decreases in absolute yield of polar fractions can be seen at high degrees of conversion. No major differences were observed between high and low temperature runs. For the four coals shown, fractions 1 and 2 increase constantly as a function of fraction 3 at both high (850°F) and low (800°F) temperatures (Figure 3-17). Fraction 4 passes through a maximum at low temperatures but not at high temperatures. Fractions 5-9 all pass through a maximum in each case, although some of the maxima are not pronounced.

All coals showed a slight tendency to produce more of fraction 9, and presumably more high-molecular weight material, at low temperatures than at high temperatures. As fraction 9 is an intermediate, this is interpreted as indicating that at high temperatures the rate of further conversion of fraction 9, relative to its formation, is greater. Fraction 7 is either never formed in significant quantities or always further converted rapidly.

At low temperatures, West Kentucky 9,14 coal gives an absolute yield of fractions 4 and 5 (most of the conventional asphaltenes) lower than that of the other three coals. At high temperatures this effect is essentially lost. Wyodak-Anderson coal always gave a product that is very rich in these fractions. Under comparable conversion conditions, at which the total SRC yield is lower for Wyodak-Anderson than for the other coals, the absolute yield of fractions 4 and 5 was actually higher than that from the other coals.

The initial products of coal liquefaction resemble the materials which can be extracted from coal directly with pyridine at low temperatures. Conversion of this residual coal after such an extraction, however, does not produce materials which are grossly different from extracts which are subjected to the same conversion conditions. Tables 3-6 and 3-7 show the similarities of such a comparison and Figure 3-18 shows that the hydrogen consumption during conversion for the extract and residue are esentially identical.

Slight differences were noted in the aromatic H content of the two SRCs, which is also reflected in the differences in the aromatic C content of the starting feeds. Very few other differences could be found, indicating that the portion of coal which can be directly extracted is chemically and structurally very similar to the non-extractable material for West Kentucky 9,14 coal.

This is not believed to be a general phenomenon, however. For example, Western coals (Wyodak-Anderson) generally have extracts much richer in hydrogen content (7%) than the residues (5%), and should behave differently on conversion.

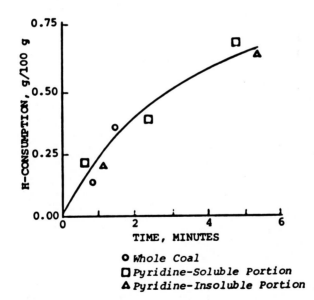

FIGURE 3-18.   Hydrogen Consumption from Solvent vs. Time for West Kentucky 9,14 Coal.

TABLE 3-6. Elemental Analyses of Pyridine-Soluble and Insoluble Feeds and Products

Feed	Soluble			Insoluble		
Sample	Feed →	SRC +	Residue	Feed →	SRC +	Residue
Treatment	PY EXT			PY EXT		
Fraction Residue	PY SOL			RES		
Percent of Coal	22.00	15.90	2.98	78.00	47.77	22.71
Percent of SRC	.00	100.00	.00	.00	100.00	.00
C	78.54	83.40	76.42	67.89	83.31	52.50
H	5.45	6.06	3.51	4.40	6.62	2.61
O	10.42	5.56	11.04	12.33	6.64	8.65
N	2.01	1.99	2.29	2.22	1.47	1.41
S	2.46	1.26	2.35	2.64	1.36	4.79
Ash	1.12	1.73	4.38	10.52	.44	30.05
Cl	.00	.00	.00	.00	.00	.00
H/C	.83	.87	.55	.78	.95	.60

Table 3-7. SESC Analyses of Pyridine-Soluble and Insoluble Feeds and Products

Feed Sample	Soluble		Insoluble
Fraction	Feed	SRC Product	SRC Product
1 HEXANE	4.05	.50	1.93
2 HEX/BZ	4.25	6.50	2.64
3 CLF	6.45	8.37	11.11
4 CLF/ET	25.60	29.85	22.37
5 ET/EOH	14.54	13.76	9.45
6 MEOH	7.89	6.76	8.69
7 CLF/OH	2.64	6.59	7.01
8 THF	10.05	2.66	13.15
9 PY	25.16	19.00	23.66

IV. THE INTERCONVERSION OF SRC FRACTIONS WITH EXTENDED REACTIONS

For a clear understanding of the chemical transformations of coal into soluble products with decreased functionality, we must understand the transformations which occur within the initially solubilized products. To accomplish this several SRCs were fractionated in large enough quantities to provide sufficient amounts (5-15 g) of key fractions so that they could be used as feedstocks. These fractions were also extensively characterized.

Two such fractions were isolated from SRCs produced in the Wilsonville pilot unit: (1) the benzene soluble fraction of West Kentucky 9,14 SRC and (2) SESC fraction 9 from Illinois #6 (Burning Star). Isolation of the benzene soluble fraction was by conventional Soxhlet extraction; it represents a mixture of SESC 2-5. The isolation of SESC fraction 9 was by sequential Soxhlet extraction of the SRC by $CHCl_3$, THF, and pyridine. The $CHCl_3$ and THF-insoluble, pyridine soluble extract constituted a concentrate of SESC-9. More detailed discussions of the conversions of these fractions have been provided elsewhere (3-6).

Fractions were also isolated from an HRI Wyodak-Anderson SRC (177-114-2B) by preparative SESC. Owing to slight differences in the preparative-scale procedure, the fractions used in these conversions do not appear to be pure by analytical SESC; rather they should be considered to be concentrations of the SESC fractions as we report them in LC analyses.

A number of general observations can be made concerning the conversion of the various SESC fractions. CO was produced in much larger amounts than CO_2 for all fractions. There appears to be a general trend of increasing tendency of char formation with increasing fraction number. The ^{1}H-NMR and elemental analyses of the various feeds and products all showed similar trends. For example, the H/C ratio of all SRC products was lower than that of the feeds. The aromatic H contents of SRCs were all higher than those of the feeds. The oxygen contents of all SRC products were lower than those of the feeds, but little correlation with hydrogen consumption was noted. It should be mentioned however, that these conversions were produced at long contact times. At short times, the products have aromatic contents similar to those of the parent coal.

A preparative SESC fractionation of several SRC products was conducted and the products further characterized. The molecular weights of the majority of the fractions were lower than those of the feedstock (SRC = 450 ± 20 and Feed = 630). A higher molecular weight component was found in fraction SESC 7 (1480 mol. wt.). Figures 3-19 show the SESC fraction distribution for the feeds and products of these conversions. Also shown are the reaction conditions and the yield of pyridine insoluble material (char) produced during the various conversions.

D. Duayne Whitehurst *et al.*

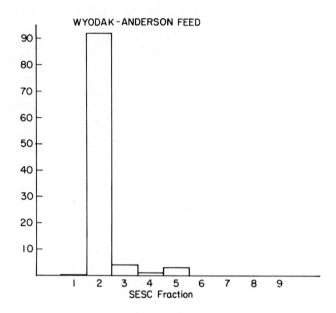

(a) Wyodak-Anderson Feed

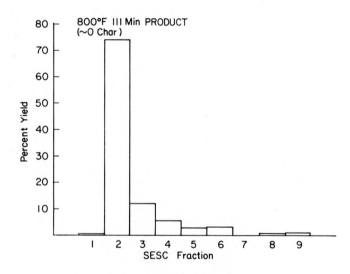

(b) 800°F 111 Min. Product

FIGURE 3-19.   SESC Fraction Distribution of Feeds and Products

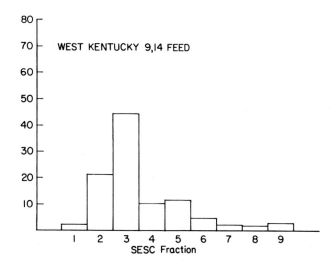

(c) West Kentucky 9,14 Feed

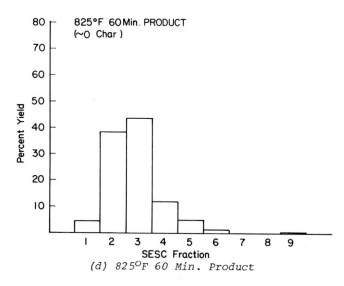

(d) 825°F 60 Min. Product

FIGURE 3-19. SESC Fraction Distribution of Feeds and Products (cont.)

D. Duayne Whitehurst *et al.*

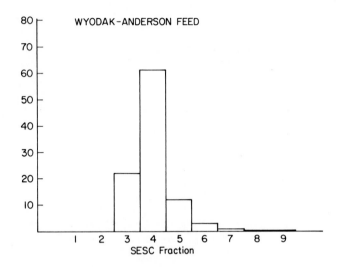

(e) *Wyodak-Anderson Feed*

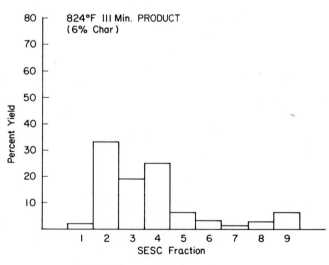

(f) *824°F 111 Min. Product*

FIGURE 3-19. *SESC Fraction Distribution of Feeds and Products (cont.)*

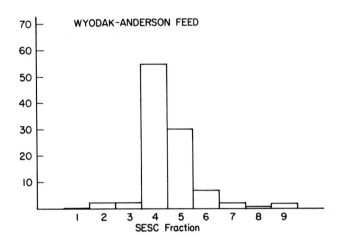

(g) Wyodak-Anderson Feed

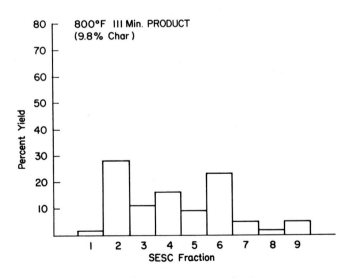

(h) 800ºF 111 Min. Product

FIGURE 3-19. SESC Fraction Distribution of Feeds and Products (cont.)

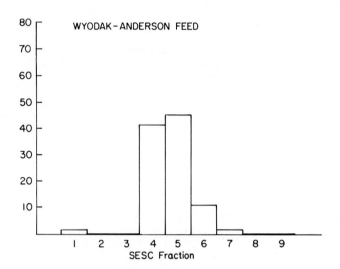

(i) Wyodak-Anderson Feed

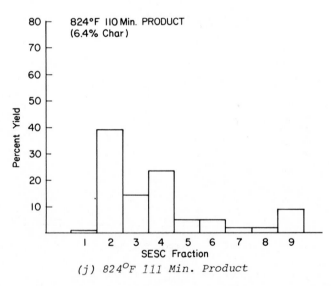

(j) 824°F 111 Min. Product

FIGURE 3-19. SESC Fraction Distribution of Feeds and Products (cont.)

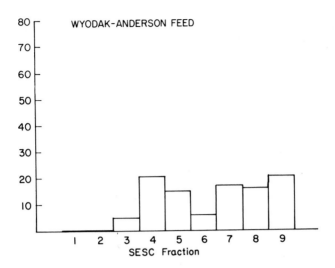

(k) Wyodak-Anderson Feed

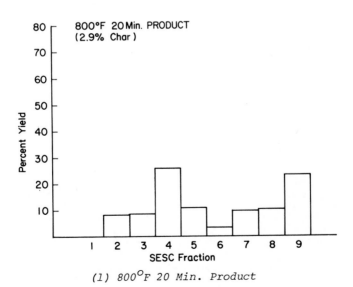

(l) 800°F 20 Min. Product

FIGURE 3-19. SESC Fraction Distribution of Feeds and Products (cont.)

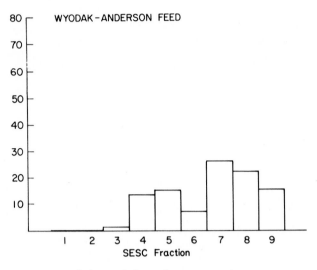

(m) Wyodak-Anderson Feed

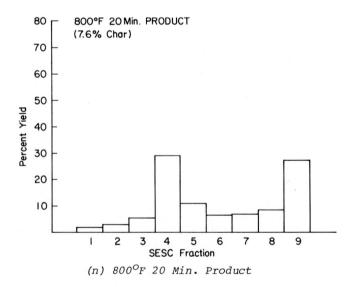

(n) 800ºF 20 Min. Product

FIGURE 3-19. SESC Fraction Distribution of Feeds and Products (cont.)

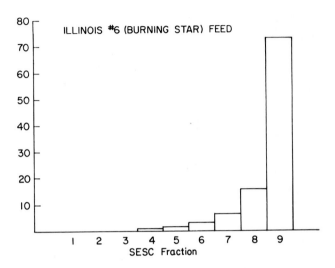

(o) Illinois #6 (Burning Star) Feed

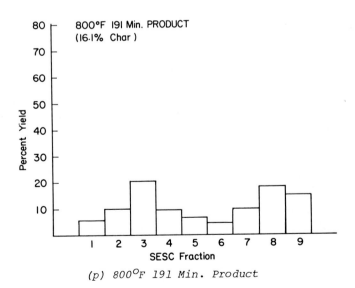

(p) 800°F 191 Min. Product

FIGURE 3-19. SESC Fraction Distribution of Feeds and Products (cont.)

These figures show that SESC-2 (aromatic hydrocarbons) is not converted to the other fractions to any significant extent. SESC-3 (heterocyclics) does produce some SESC-2, though the conversions were not extensive.

The mono and diphenols and basic nitrogen heterocyclics (SESC-4 and 5) underwent significant transformations. Both lower and higher functionality products were produced though lower functionality fractions (SESC-2 and 3) were predominant.

Polyfunctional components of SRCs (SESC-6—9) were found to be significantly more reactive than less functional components. On reaction, they produced primarily lower functionality products (presumably via dehydroxylation). For bituminous coals SESC-3 was the predominant product. For subbituminous coals SESC-4 was the major product.

These data indicate that a detailed kinetic description of the transformation of coal products during liquefaction will be very difficult. One major complication is condensation reactions between coal liquid fractions and phenols in the solvent. Such condensations can result in the production of products which are more highly functional than the starting materials. The formation of char is also a competing side reaction.

V.   THE REMOVAL OF OXYGEN AND SULFUR FROM SRC

Under standard conditions (800°F, 1000 psi H_2, in synthetic solvent), the sulfur content of the SRC was found to be linearly related to the oxygen content of the SRC. The data for Illinois #6 (Monterey) and Illinois #6 (Burning Star) coals as well as Wyodak-Anderson and West Kentucky 9,14 coals are presented in Figures 3-20(a)-(d). SRCs obtained at high temperatures (840-880°F) produced much more scatter in the data, with no apparent trend.

The linear correlation appears quite good (least squares fit) for Illinois #6 (Burning Star) and Wyodak-Anderson coals with somewhat more scatter for the other coals. An interesting observation is noted in Figure 3-20(d). Here the S/O line was forced through the origin with the best least squares fit for the rest of the data points. Wyodak-Anderson and Illinois #6 (Burning Star) data points fit exceptionally well and Illinois #6 (Monterey) and West Kentucky 9,14 coals showed no more scatter than in unforced fits (Figures 3-20(b) and (c)). This correlation indicates that at all levels of oxygen and sulfur in the SRC, the rate-limiting steps for oxygen and sulfur removal are the same in a thermal hydrogen donation process. Thus, the rate constants for deoxygenation and desulfurization must be approximately equal. As

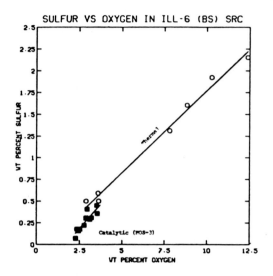

3-20 (a). Sulfur vs. Oxygen in Illinois #6 (BS) SRC

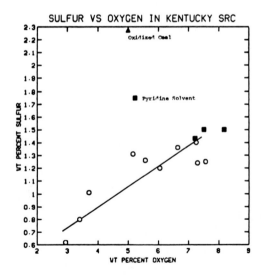

3-20 (b). Sulfur vs. Oxygen in Kentucky SRC

FIGURE 3-20. Sulfur vs. Oxygen in SRCs

D. Duayne Whitehurst *et al.*

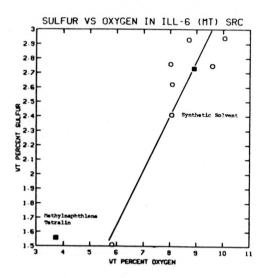

3-20(c). Sulfur vs. Oxygen in Illinois #6 (MT) SRC

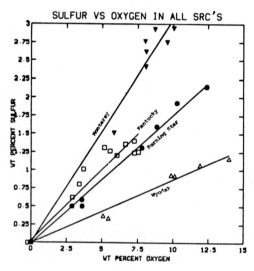

3-20(d). Sulfur vs. Oxygen in All SRCs

FIGURE 3-20.   Sulfur vs. Oxygen in SRCs (cont.)

will be described later, the phenol content (meg/gm) of the SRC
changed by a factor of two with short and long contact times in
the Illinois #6 (Monterey) SRC.

These rate constants are perturbed by the absence of hydrogen
donors. Using pyridine as the solvent (long time), significantly
less desulfurization for a given level of eoxygenation was
achieved. A similar observation was noted for Illinois #6
(Burning Star) coal in naphthalene-biphenyl solvent. Oxidation of
the coal also produced a different curve which had unusually high
sulfur. Additional studies showed that high concentrations of
hydroaromatics gave high rates of desulfurization.

The relative rates of sulfur and oxygen removal can be altered
by several factors. One commonly practiced technique is the use
of catalysts. For comparative purposes all of the coal liquids
hydrogenation data reported (3-16) by S. E. Voltz et al., of MRDS
(Paulsboro Laboratory) (EPRI 361-1), for Illinois #6 (Burning
Star) SRC and SRC plus solvent at 800°F were plotted in a similar
manner. Figure 3-20(a) shows a direct comparison of catalytic and
thermal processes. The catalyst used for these data was
commercial HDS-3. Other catalysts produced different curves which
are consistent with the reported varying desulfurization/
deoxygenation activities. The data in Figure 3-20(a) clearly show
a change in selectivity for the catalytic process, which is more
efficient for sulfur removal.

L. D. Rollmann has reported (3-17, 3-18) some catalytic
studies pertinent to the above discussions. In studies of
competitive deoxygenation and desulfurization of a wide variety
of model compounds, he found the following relative rates of
hydrogenation and/or hydrogenolysis.

In addition, he reported that at low temperatures (650°F)
desulfurization is extremely selective toward hydrogenolysis
without ring saturation and deoxygenation of monophenols or
heterocyclic ethers requires ring saturation prior to oxygen
removal. At high temperatures (750°F) this selectivity is less
pronounced.

$$\xrightarrow{<750^\circ F} \quad 2 \bigcirc \quad + \quad H_2O\,(6H_2)$$

$$\xrightarrow{<750^\circ F} \quad 2 \bigcirc \quad + \quad H_2S\,(2H_2)$$

$$\xrightarrow{<750^\circ F} \quad + \quad H_2O\,(6H_2)$$

$$\xrightarrow{<750^\circ F} \quad + \quad H_2S\,(3H_2)$$

These observations are consistent with the specific catalytic desulfurization mechanism proposed by R. J. Mikovsky (3-19), or other mechanisms recently reviewed by V. H. J. DeBeer and G. C. A. Schuit (3-20).

The structural studies reported earlier on Illinois #6 (Monterey) SRC indicate that this SRC contains oxygen and sulfur substantially in heterocyclic ring structures. Thus, catalytic hydrodesulfurization at low temperature would appear the most promising method for selective removal with lower hydrogen consumption.

VI. RELATIONSHIPS BETWEEN COAL LIQUEFACTION BEHAVIOR AND COAL STRUCTURE

One general question on the structure of coal is whether Eastern and Western coals are similar in composition as their ages differ by 60 million years. A structural comparison was conducted using the SRC of a Hiawatha coal from Utah (705°F, 1500 psi H_2, 31 min.), supplied through the courtesy of L. Anderson, of the University of Utah. This material was fractionated and analyzed and compared to the Eastern bituminous coals we have previously studied (West Kentucky 9,14, Illinois #6.)

The analyses shown in Table 3-8 indicate major compositional differences between this SRC and SRCs from Eastern bituminous coals. In particular, unusually high contents of fractions SESC-1 and 2 are observed. The content of saturated and aromatic hydrocarbons in an Eastern bituminous SRC with a similar distribution of fraction 3-9 would be .5% of fraction 1 and 10% of fraction 2, as compared to 5.3% and 23% respectively in this SRC. The most likely explanation of this difference is the initial presence of such naturally-occurring hydrocarbons in the initial coal.

TABLE 3-8. Utah 375 SRC

Fraction Number	Percent in SRC	Elemental Analysis				Molecular Weight	H/C	Percent Aromatic H	Percent Aromatic C
		% C	% H	% O	% N				
1	5.3	88.0	12.0	-	-	-	1.6	-	-
2	23.0	93.3	6.4	-	-	-	0.82	58	66
3	11.0	87.9	7.0	3.4	1.0	395	0.96	36	62
4	13.7	84.9	6.9	6.1	1.9	413	0.98	40	63
5	3.3	-	-	-	-	-	-	-	-
6	3.0	-	-	-	-	-	-	-	-
7-9	40.7	-	-	-	-	-	-	-	-
All SRC	100	82.4	6.6	3.8	1.3	-	0.95	-	58
Coal	-	81.3	5.7	10.3	2.0	-	0.84	-	-

We have estimated the average molecular structure of SESC-3 of this SRC and SRCs of other coals as shown in Figure 3-14. This figure also gives some additional compositional data. The skeletal C-H structure of Hiawatha SRC is between that of West Kentucky 9,14 and Illinois #6 (Monterey) coal. Indeed, Hiawatha coal is a very reactive coal and easy to solubilize like West Kentucky 9,14, but has a significantly lower sulfur content (0.7%).

These data suggest that Western bituminous coals are not grossly different in skeletal structure from Eastern bituminous coals; however, they do produce more hydrocarbons and, in this case, yield a low sulfur SRC.

One more comment could be of interest. The data in Table 3-8 indicate that the H-consumption for liquefaction of different coals increases with an increase in the number of polycondensed saturated rings in their structure. The increase in H-consumption also parallels an increase in oxygen content in coals. However, considering the high H-consumption during further conversion of fractions SESC-2, 3, 4 and 5 in Wyodak-Anderson coal, all with rather low heteroatom content, the influence of the C-H skeleton of coal seems to be an equally important factor in the H-consumption during liquefaction.

REFERENCES

3-1.   M. Farcasiu, FUEL, 56, 9 (1977).
3-2.   D. D. Whitehurst, M. Farcasiu, T. O. Mitchell, and J. J.
        Dickert, Jr., "The Nature and Origin of Asphaltenes in
        Processed Coals", EPRI Report AF-480, Second Annual Report
        Under Project RP-410-1, July 1977.
3-3.a. P. H. Given, D. C. Cronauer, W. Spackman, H. L. Lovell,
        A. Davis and B. Biswas, FUEL 1975, 54, 34.
3-3.b. P. H. Given, D. C. Cronauer, W. Spackman, H. L. Lovell,
        A. Davis and B. Biswas, FUEL 1975, 54, 40.
3-4.   M. Farcasiu, T. O. Mitchell, and D. D. Whitehurst,
        "Asphaltols - Keys to Coal Liquefaction", CHEMTECH, 7,
        November 1977, pp. 680-686.
3-5.   B. L. Karger, L. R. Snyder, and C. Horvath, An Introduction
        to Separation Science, Wiley, New York, 1973.
3-6.a. D. D. Whitehurst, M. Farcasiu, and T. O. Mitchell, "The
        Nature and Origin of Asphaltenes in Processed Coals",
        EPRI Report AF-252, First Annual Report Under Project
        RP-410, February 1976.
3-6.b. T. O. Mitchell and D. D. Whitehurst, "Short Contact Time
        Coal Liquefaction", ACS Division of Fuel Chemistry,
        Reprints Page 127, San Francisco, August 1976.
3-7.   R. C. Neavel, FUEL, 55, 237 (1976).
3-8.   J. K. Brown and W. R. Ladner, FUEL, 39, 79 (1960).
3-9.   J. K. Brown, W. R. Ladner and H. Sheppard, FUEL, 39, 87
        (1960).
3-10.   K. T. Bartle, T. G. Martin and D. K. Williams, FUEL, 54,
        226 (1975).
3-11.a. A concise introduction to the subject of polarography can
        be found in "Polarography", by D. R. Crow, Methuen, London,
        (1968). The derivation of the Ilkovic equation appears in
        the appendix.
3-11.b. I. Koltoff, and J. J. Lingane, "Polarography", Vol. 2, 2nd
        Edition, Interscience Publ., New York, 1952, p. 638.
3-12.   E. Z. Leete, J. Am. Chem. Soc., 83, 3645 (1961).
3-13.   N. L. Allinger, M. T. Tribble, M. A. Miller and D. H. Wertz,
        JACS, 93, 1637 (1971).
3-14.   E. M. Engler, J. D. Amdose and P. V. R. Schleyer, JACS, 95,
        8005 (1973).
3-15.   M. Farcasiu, T. O. Mitchell, and D. D. Whitehurst, Pro-
        ceedings of Coal Chemistry Workshop, SRI, August 1976.
3-16.   J. G. Bendoraitis, A. V. Cabal, R. B. Callen, T. R. Stein,
        and S. E. Voltz, "Upgrading of Coal Liquids for Use as
        Power Generation Fuels", EPRI Project 361-1, Phase I
        Report, January 1976.

3-17. L. D. Rollmann, ACS Preprints, Division of Fuel Chemistry, 21, 59 (1976).

3-18. L. D. Rollmann, J. Catalysis, in press.

3-19. R. J. Mikovsky, A. J. Silvestri and H. Heinemann, J. Catalysis, 34, 324 (1974).

3-20. V. H. J. deBeer and G. C. A. Schuit, Annals of the New York Academy of Sciences, 272, 61 (1976).

3-21. D. D. Whitehurst, T. O. Mitchell, M. Farcasiu, J. J. Dickert, Jr., "The Nature and Origin of Asphaltenes in Processed Coals", Final Report to EPRI Under Project RP-410, in press.

Chapter 4

SIGNIFICANCE OF THE PHYSICAL PROPERTIES

OF COAL TO COAL CONVERSIONS

In the initial phases of conversion, the physical properties of coals may play an important role in controlling the rate of coal solubilization, the rate of diffusion of reagents into the coal matrix and the rate of diffusion of coal products out of the matrix.

This chapter addresses the physical mechanisms in coal dissolution and the significance of mass transport and diffusion phenomena to the rate of conversion, product distribution, and product yield. The results will be discussed in terms of how coal liquefaction behavior is altered on changing the physical properties of a coal. The effects of physical property variation can be studied by essentially the following two basic experimental procedures: (1) varying the properties of the feed coal (by particle size changes and thermal or chemical pretreatments) and (2) examining the physical properties of liquefaction residues, especially those obtained under mild conditions. A key question is the extent to which particle integrity is maintained as coals are converted.

The following major topics in this chapter are discussed in terms of their significance to coal liquefaction.

- Surface areas and pore systems of coals
- Solvent swelling of coals and selective sorption of solvent components
- Coal particle size
- Heating and oxidation pretreatments

I.   SURFACE AREA AND POROSITY

It has been proposed that the relative reactivity of coal
correlates with its surface area and that the surface area changes
in a systematic manner during its conversion to soluble form (4-1).
Determining the importance of such physical factors requires
measurements of density, particle density, average pore diameter,
N_2 and CO_2 surface area, pore volume, and pore size distribution
(the fraction of total pore volume as a function of pore diameter).
However, measuring these parameters is complicated since coals are
reactive materials, sensitive to drying and degassing procedures
and to interacting with adsorbates. In addition, they are
thermally unstable, contain volatile materials and may swell on
heating. Another complication is that coal samples are often
available only as fine powders; residues of reactions invariably
are fine powders. Since the primary interest is in the structure
of a solid in a slurry at greater than 800°F, formidable experimental and interpretation problems are encountered.

Preparation of a coal or residue for sorption or displacement
measurements requires that the sample be dried and degassed, but
the severity of the degassing step can have a significant influence
on the subsequently determined surface area, pore volume and pore
size distribution. Even in the drying step (typically heating in
air or an inert atmosphere to a little over 212°F) reactions may
occur, especially for low-ranked coals. For example, lignites
that have been dried thoroughly will not absorb as much water as
they originally contained, which has been attributed to an
irreversible alteration of a gel-like structure (4-1).

For caking and swelling coals, there is the additional problem
that the gross morphology of the sample will change on heating,
especially in the presence of certain solvents. A highly caking
coal will triple in CO_2 surface area if exposed to pyridine vapors
overnight.

An increase in surface area with increased degassing
temperature of air-dried samples can be attributed to breaking
down of surface oxidation complexes as the temperature is increased.
Reversibly chemisorbed oxygen may either block off small pores or
occupy sites that otherwise would be covered by N_2. Of course,
if the coal is dried or degassed at high temperatures, volatile
matter may be driven off and if air is present, oxidation reactions
will occur.

Determining the volume of pores which are too large to detect
by N_2 absorption presents additional problems since such large
pores may have dimensions of the same order of magnitude as the
interparticle voids in fine powders. In the Hg porosimetry
technique it is often not possible to distinguish between
interparticle and intraparticle voids. Concentrations of large-

pore volume which appear high may actually be due to a small
average particle size.

Large pores can also be examined by transmission electron
microscopy for information on shape, size distribution, extent of
interconnection, and other parameters. Surface areas can be
evaluated by small-angle x-ray scattering. In addition to sorption
of N_2, CO_2, and Hg, other materials such as methanol, can be sorbed
to give information on the distribution of pore volume and surface
in micropores.

Measurements of N_2 sorption and Hg porosimetry for a representative series of coals from lignite to bituminous after several
pretreatments are provided in Table 4-1. In Table 4-2 similar
data are shown for some conversion residues. The properties
measured this way may not be exactly the same under liquefaction
conditions.

The following conclusions have been drawn from these measurements in combination with other available sorption data.

First, all coals and residues have low pore volumes,
especially in the important region of 7 to 300 Å radius. In one
reported study (4-1) of 12 coals from anthracite to lignite, 11
had total pore volumes less than 16 cc/100 g; one of the three high
volatile C bituminous coals had a pore volume of 23 cc/ 100 g.
The coals discussed in this chapter have total pore volumes from
8 to 16 cc/100 g.

Second, exhaustive extraction of West Kentucky 9,14 (bituminous) coal with pyridine increases the pore volumes, including the
range of 7-300 Å radius where diffusion could be critical. The
magnitude of the pore volumes, however, remains very low. Extraction of Wyodak-Anderson (subbituminous) coal shows little effect,
and only 11% of the coal is removed.

Third, under liquefaction conditions, as dissolution progresses, the pore volumes of West Kentucky 9,14 coal in the small
and middle size ranges increase, but still do not exceed a total
of 7 cc/100 g. We have no reliable measurements of macropores for
the reason given above. Especially when the particles retain
their integrity and do not break up, the macropore volume must be
substantial at high conversions. This has been demonstrated in an
unstirred reactor (4-2,4-3,4-4).

Fourth, the CO_2 surface areas increase dramatically from 100
to 300 m^2/g on pyridine extraction of West Kentucky 9,14 coal, and
change little on subsequent liquefaction. The N_2 surface areas of
residues go through a maximum value and can be as large as 30-40
m^2/g depending on level of conversion.

High-temperature treatment in pyridine followed by extraction
gives little or no increased dissolution. However, the insoluble
residue yielded the largest observed increases in surface area.
This may well be indicative of swelling. Residues which result

D. Duayne Whitehurst *et al.*

TABLE 4-1   COAL PORE VOLUMES, cc/g, AND REAL DENSITIES (He) g/cc

1. Pretreatment: $350^{\circ}C$ in air 1 hr, then $400^{\circ}C$ vacuum 0.5 hr

Coals	Pore Volume cc/100 g			He Density
	$<7\overset{\circ}{A}$	$7-300\overset{\circ}{A}$	$>300\overset{\circ}{A}$[a]	
Pittsburgh	0.012	0.039	0.350	–
West Kentucky	0.013	0.034	0.226	–
Illinois (Burning Star)	0.000	0.041	0.378	–
N. D. Lignite	0.000	0.006	0.258	–
Wyodak	0.000	0.012	0.355	–
West Kentucky ($400^{\circ}C$ vac only)	0.000	0.002	–	1.39

2. Pretreatment: $150^{\circ}C$ in vacuum, 16 hr

Coals				He Density
Pittsburgh	0.000	0.001	–	1.39
West Kentucky	0.001	0.003	–	1.36
Illinois (Burning Star)	0.000	0.024	–	1.34
N. D. Lignite	0.002	0.000	–	1.42
Wyodak	0.000	0.012	–	1.32

3. Pretreatment: $204^{\circ}C$ flowing N_2, 16 hr

Coals			
Pittsburgh	0.000	0.001	–
West Kentucky	0.000	0.004	–
Illinois (Burning Star)	0.000	0.014	
N. D. Lignite	0.000	0.001	–
Wyodak	0.000	0.007	–

4. Pretreatment: $105^{\circ}C$ in air for 1 hr, then $130^{\circ}C$ in vacuum, 12 hr (40-70 mesh coals) (Estimates from Reference 4-1, see discussion)

Coals			
Pittsburgh	0.02	0.00	0.02
West Kentucky	0.04	0.03	0.02
Illinois (Burning Star)	0.07	0.06	0.03
N. D. Lignite	0.03	0.00	0.07
Wyodak	0.07	0.01	0.02

[a]Hg porosimetry questionable for powders, see discussion.

TABLE 4-2.   Residue Pore Volumes, cc/100 g, and
             Real Densities (He) (g/cc)

		Pore Volume cc/100 g			Real Density
	Conversion	<7Å	7-300Å	>300Å a	g/cc

1. Pretreatment: 350°C in air 1 hr, then 400°C vac 0.5 hr

Pyridine Extracted

West Kentucky	(31)	0.0	1.3	24.9	1.50
AC-10	(50)	0.0	2.6	38.6	1.57
AC-14	(70)	0.0	2.9	59.2	1.58
AC-9	(78)	0.1	4.0	50.5	1.70
AC-7	(93)	0.6	6.1	24.2	2.02
J7939	(95)	0.0	0.5	38.6	-
J7939 (400°C vac only)		0.1	0.1	12.5	1.86

2. Pretreatment: 150°C in vacuum, 16 hr

Pyridine Extracted

West Kentucky	(31)	0.1	0.4	5.7	1.32
AC-3	(27)	0.2	0.1	-	1.53
AC-10	(50)	0.2	1.6	0.8	1.27
AC-14	(70)	0.3	0.4	46.7	1.54
AC-16	(57)	0.1	0.3	16.4	1.63
AC-17	(29)	1.0	2.3	-	1.44
AC-18	(38)	0.4	1.5	64.9	1.28
Pyridine Extracted					
Wyodak	(11)	0.1	0.1	12.1	1.38
AC-19	(39)	0.1	0.3	-	1.42
J7939	(95)	0.1	0.1	14.5	1.99

3. Pretreatment: 204°C flowing N_2, 16 hr

Pyridine Extracted

West Kentucky		0.3	0.6	-	-
AC-3		1.0	0.6	-	-
AC-16		0.6	2.0	-	-
AC-17		1.0	0.0	-	-
AC-18		1.0	3.0	-	-
AC-19		0.2	1.9	-	-
J7939		0.2	0.1	-	-

a Not accurate for powders, see discussion.

from coal conversion in hydrogen donor solvents give increasing
surface area with increasing conversion. Exposure of the mineral
matter may complicate interpretations however as mineral matter
from low temperature ashing of coals has surface areas on the
order of 10 m^2/g. Thus at 30-40 m^2/g the carbonaceous portion
of high conversion residues may have a surface area somewhat higher than the bulk.

One notable observation in this study is that significant
changes in pores of less than 300 Å radius, as well as large increases in CO_2 surface area, may occur by merely extracting coal
with pyridine (see Tables 4-1 and 4-2).

Two types of pore systems in coal thus appear significant.
One is associated with materials which are directly extractable by
pyridine at low temperature and contributes to the pore system of
as-received coals. These materials may be present as a "coating"
on the skeletal structure of non-extractable coal; their removal
would then leave a different type of pore system. Alternatively,
the pyridine extractable portion of the coal could be present as
discrete regions of the coal particles such as a specific maceral.
The other pore system is that of the pyridine insoluble portion of
the coal.

The lower range of the pore size distributions which are
normally reported for coals may in fact be affected by the presence
of extractable material, and may not be indicative of the pore
systems that are critical at high levels of coal conversion.

The effects on mass transport limitations can not be quanti-
fied at present. It may be reasonable, however, to expect that
the migration of the soluble material from within the pore system
of the pyridine-insoluble matrix (which maintains its integrity
at low conversion) to the exterior solution may have diffusional
restrictions. If self-condensation of soluble material occurs,
forming char within the pores, this effect may be accentuated.

The physical properties of residues at severe conditions have
been reported. One paper (4-5) examined residue N_2 surface areas
for long autoclave runs. For 800°F runs, the surface area increased to 30 m^2/g at 45 min, and then gradually decreased to
5.6 m^2/g at 300 min. This decrease was attributed to the initial
residue's pore system being filled with insoluble material (char)
formed from the initial liquefied product. Examination of coal
liquefaction residues by optical microscopy (4-6,4-7) sometimes
shows anisotropic carbonaceous deposits surrounding inorganic or
unreacted carbonaceous cores. Such coatings are believed to have
been formed from initially solubilized materials. Chapter 8,
"Char Formation", discusses the formation of solids in more detail.

Based on these and other data the following picture emerges.
Coal particles have very little micropore structure (less than
300 Å) low total pore volume, and low surface area. They can be

envisioned as fairly solid blocks with perhaps up to 5-10% of the
volume consisting of voids. Very roughly, a third of the void
volume is too small to admit organic molecules, and an additional
third of the pore system could admit solvent molecules and low
molecular weight product molecules only. As dissolution progresses,
the contributions of these two pore size ranges increase only
slightly, suggesting that they do not play an important role in
the dissolution process. In addition, much of the initial product
may have a molecular weight greater than 1000; most of the dissolution is complete within 2 minutes, in which time large molecules could not escape unless there were a substantial network
of very large pores. Tentatively it is proposed that the dissolution may take place substantially by a shell progressive
mechanism, although concurrent reactions are occurring within the
particles.

II. CHANGES IN GROSS PARTICLE MORPHOLOGY WITH
     DEGREE OF CONVERSION

     It is important to know how the overall size and morphology
of coal and residue particles change when they are subjected to
various treatments (heating, oxidation, extraction, and liquefaction under stirred and unstirred conditions). Solvent swelling
of coals, sorption by coals, and photomicrographs of a number of
coals and residues have been examined.

A. Solvent Swelling of Coal: It is well known that coals will
swell when immersed in organic solvents (4-8). The extent of this
expansion varies with both the rank of coal and the specific solvent. Linear expansions of up to 5% have been found (4-8) and
amines generally produce the largest changes (4-9). High heats of
wetting are associated with solvents which swell coals most significantly. Since this property may give insight into the interaction of solvent with coal, the equilibrium swelling of West
Kentucky 9,14 coal by several solvents, using a duPont 990 Thermo-
mechanical Analyzer has been investigated (4-2). Some results
are presented below:

SOLVENT SWELLING OF WEST KENTUCKY 9,14 COAL

Solvent	% Increase in Volume (212°F)
γ-Picoline	1.0
Synthetic Solvent*	0.8
Process-Derived Recycle Solvent	0.2

*43% Tetralin, 18% p-cresol, 2% γ-picoline, 37% methylnaphthalene

The optimal procedure for this measurement has not yet been
determined, but the relative values of the results are believed to
be meaningful. The percent volume increase is defined, assuming
the bulk density of coal to be 1.0, as

$$\% \text{ Volume Increase} = \frac{V(\text{final}) - V(\text{original})}{V(\text{original})}$$

As can be seen, the synthetic recycle mixture interacts quite
strongly with coal -- almost as well as γ-picoline and somewhat
more than a process-derived recycle solvent. This indicates that
the interaction of the synthetic solvent (used extensively in work
by MRDC reported in this book) with the coal matrix should be
similar to that of an actual recycle solvent.

B.   Selective Sorptions by Coal: Gas chromatographic analyses of
the solvent portion of 1:1 slurries of coal in synthetic solvents
were obtained. A West Kentucky 9,14 coal slurry had been stored
for 10 weeks; Wyodak-Anderson slurry had been stored for 10 days.
The results are given below with the solvent analyses for comparison:

Component	Solvent #5	West Kentucky Slurry in #5	Solvent #6	Wyodak Slurry in #6
γ-Picoline	1.91	0.43	2.00	0.82
p-Cresol	16.62	7.71	17.31	8.12
Methylindan	1.13	0	0.59	0.50
Tetralin	40.09	48.94	42.80	48.70
Naphthalene	1.06	2.48	0	0
Methylnaphthalene	38.19	40.44	37.31	41.82

It can be seen that both coals readily sorb γ-picoline and
p-cresol selectively, and to a lesser extent methylnaphthalene.
When such a coal/solvent slurry is injected into an autoclave, the solvent within the pores would be particularly rich
in cresol. On subjecting the coal which contained sorbed p-cresol
to liquefaction conditions for only a few minutes, however, the
external solvent composition was restored. It therefore appears
that either (a) extremely fast equilibration of these small
molecules occurs after injection at reaction conditions, and

residues do not selectively absorb polar and aromatic materials or
(b) these solvent molecules are displaced from within the particles
by reaction products.

Another possible factor is that in a rapidly-stirred reactor
coal particles would be broken up to such an extent that the internal pore volume becomes insignificant. Particles do disintegrate rapidly during fast-stirred reactions, but not during unstirred reactions (as will be shown later in this chapter).

C. Photomicroscopy: Perhaps the most definitive tool for evaluating the changes in morphology of coal particles during coal
conversion is photomicroscopy. Using this technique, it has been
reported (4-10) that coal particles break up after very short time
exposures to hydrogen donor solvents in a stirred autoclave. Particles were observed to break up substantially in tetralin at
680°F in experiments with "immediate cooling" after heat-up. Since
the extent of exposure was not clearly defined, it is difficult
to quantify these experiments. It was concluded, however, that
"disintegration of coal particles in a hydrogen-donor-solvent
occurred almost instantaneously at 680°F".

The reported experiments (4-10,4-11) were conducted in an
autoclave stirred at speeds in excess of 500 rpm (the agitator was
slowed to 500 rpm for sampling). The same authors (4-12) have
found that stirring rate affects the breakage of the particles.
Hydrogen-donor-solvent attack was proposed to aid the weakening
process, since this break-up was not observed in experiments with
paraffin oil, a non-hydrogen-donor solvent (4-10,4-11). The
possibility of a reversible recondensation of intermediates was
not considered, however.

The stresses on the coal particles in a stirred autoclave are
much more severe than they might be in some envisioned commercial
liquefaction processes. Evaluations of possible mass transport
limitations in such systems require an understanding of the gross
morphology of particles and the changes that may occur during dissolution; our investigations are described below (4-2,4-28).

In the early stages of dissolution, using tetralin (instead
of a synthetic solvent mixture), has little significant effect on
coal conversion to pyridine-soluble material. At less than 30-
seconds reaction time, there is no significant difference in conversion between the stirred and unstirred reactors. In some runs
coal was injected dry rather than as a slurry, but the results
were similar, showing that this also is not significant to the
initial phases of the process.

The first series of unstirred tests involved injecting either
dry coal or slurries into hot solvent, allowing a short reaction
time, and then quenching the reaction by plunging the entire
reactor into an ice-water bath. Ash contents of the original coal
and reaction residues were used to calculate the conversions.

Figure 4-1 shows photomicrographs of West Kentucky 9,14 coal
(a) before reaction, (b) and (c) as recovered after reactions and
(d) after applying slight pressure to the skeletal fragments after
reaction. Overall size and shape of the particles after reaction
was not changed in the absence of stirring. The edges were, however, more jagged, suggesting a chemical attack on the particles.
The particles shown (Figure 4-1) are the residue after 85-87%
conversion. Residual particles after lesser degrees of conversion
show a great similarity to the original coal particles, as may be
expected. These results clearly demonstrate that the physical
integrity of the coal particle is not necessarily lost on contact
with hot solvent followed by a rapid quench.

The result of mild pressure on the skeletal structure is
shown in Figure 4-1(d). It is not surprising that the residual
structure is fragile, since only 15% of the original organic
material is left, along with the ash, in a structure of approximately the same outer dimensions as the original coal particle.
The unstirred reactor did not apply mechanical stresses to the
particles, thus allowing them to remain intact in spite of a
large fraction of material having been removed. In Figures 4-2(b)
and 4-3(a & b) are shown additional photomicrographs of West
Kentucky 9,14 coal run in an unstirred reactor under various conditions of temperature, time, and solvent. Yields of pyridine
soluble material ranged from 69% to 87%. In all cases the particles retained their integrity. In fact, there is an indication
that some of the residue particles may be larger than the original
coal particles. (Compare Figures 4-2(a) and 4-2(b).) Solvent
swelling and/or thermal expansion may have been observed.

Figure 4-2(c), however, shows a sample from the stirred autoclave at only 52% conversion, and it can be seen that the particle
size has been drastically reduced after 120 seconds.

Figure 4-4 shows scanning electron micrographs of West
Kentucky 9,14 coal before and after Soxhlet pyridine extraction.
The parent coal (a) appears to be quite featureless; there are no
apparent cracks or openings (that would have to be about 5-10μ to
be clearly seen). The surface is covered with small particles
that could be either fine hydrocarbon "dust" or mineral matter
that is either resting on or embedded in the surface. Figure
4-4(b) shows several of these fine particles, while 4-4(c) shows
a single extracted coal particle with about 30% of the organic
material removed. It is larger than the parent particles and is
clearly cracked and pitted; it shows no "dust" particles on its
surface. Figure 4-4(d) shows this particle at a higher magnification.

The retention of particle integrity of a subbituminous
(Wyodak-Anderson) coal in the absence of stirring has been evaluated in an experiment at 820°F in tetralin for 2 minutes in an

(a)   Original coal 212-600µ
particles.

(b)   After 2 min. in
tetralin, unstirred, at 832°F;
86% conversion.

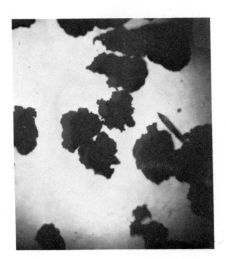

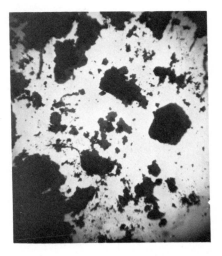

(c)   After 1 min. in tetralin at 850°F; 86% conversion.

(d)   Result of applying
slight pressure to particles
shown in (c).

FIGURE 4-1.   Photomicrographs of West Kentucky Coal and Residues.

PARENT COAL

AFTER CONVERSION (~ 80 %)
2 MIN. 425 °C NO STIRRING

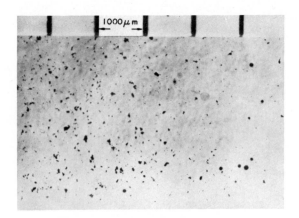

AFTER CONVERSION (80%)
2 MIN. 425 °C RAPID STIRRING

FIGURE 4-2.   *West Kentucky Coal Particles.*

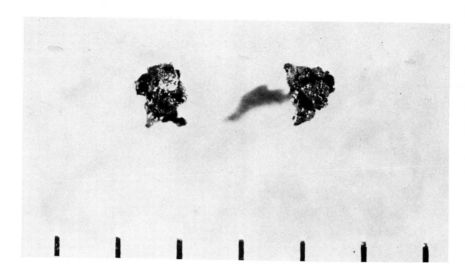

(a)

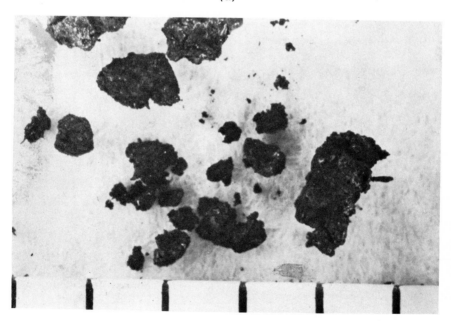

(b)

FIGURE 4-3.   West Kentucky Coal Conversion Residues.   Photomicrographs of particles from conversion in unstirred reactor.
(a)   Initially 600-212μ particles run in tetralin at 800°F for
60 sec.; 69% conversion.   (b)   Initially 600-212μ particles run
in synthetic solvent at 824°F for 60 sec.; 78% conversion.
1000μ between scale lines.

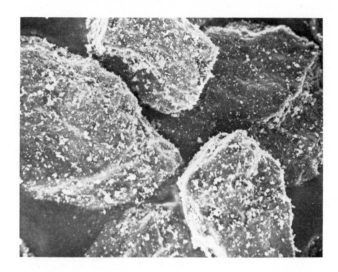

(a)   *Scanning election micrograph of 212-600µ particles
of West Kentucky Coal.   100X   ∿10µ/mm*

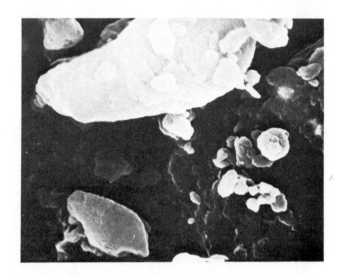

(b)   *Same particle 5025X ∿0.20µ/mm.*

*FIGURE 4-4.   Scanning Electron Micrographs of West Kentucky
Coal and Pyridine Extraction Residue.*

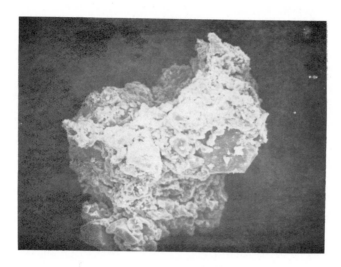

(c) *Scanning electron micrograph of a single 400μ particle of West Kentucky Coal after pyridine extraction. X100 7.5μ/mm.*

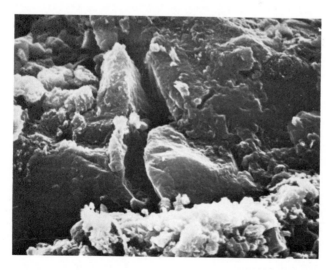

(d) *Same particle 5025X ∿0.20μ/mm.*

FIGURE 4-4. Scanning Electron Micrographs of West Kentucky Coal and Pyridine Residue.

unstirred reactor. Figure 4-5 shows photomicrographs of the Wyodak-Anderson coal before reaction and of the reaction residue. In comparing this result with the observations on the West Kentucky 9,14 coal, several similarities and several differences may be noted: (a) the Wyodak-Anderson residue particles did not disintegrate in spite of the high-temperature and substantial conversion (about 70%); (b) the Wyodak-Anderson residue particles are slightly smaller than the parent coal; the West Kentucky 9,14 residue particles were larger, suggesting swelling during reaction; (c) the Wyodak-Anderson residue appears to have a layered-type porous structure, whereas the West Kentucky 9,14 residue had a cellular-type porous structure, and (d) the residue of each was fragile and easily shattered.

Photomicrographs of the residue from a stirred reactor run with 425-600µ Wyodak-Anderson coal(in a donor solvent at 810°F for 120 seconds)showed that the residue particles were extremely small. This is in line with previous observations for West Kentucky 9,14 coal. Thus rapid stirring during coal dissolution can provide enough stress to completely disintegrate particles of both bituminous and subbituminous coals.

In a commercial process,the reactor turbulence might be somewhat less severe so that the retention of particle integrity could be somewhere between these two extremes. Therefore, several experiments were conducted with intermediate reactor turbulence.

In a run with 25-600µ particles of Wyodak-Anderson coal, (for 120 seconds, at 806°F in synthetic solvent with 63% conversion in a stirred reactor) the stirrer was operated by hand at only about 120 rpm. Even in this case, the particles had disintegrated.

Particles do survive injection procedures and at least up to 33% conversion (15% more than the initially extractable material) they can survive even a fast quench procedure. However, at 63% conversion they disintegrate as a result of either gentle stirring during a run or turbulence during the quench. Thus, it is not very likely that particles could survive in a commercial SRC unit past 50% conversion, although we do not know the exact stresses to which they would be subjected in such units.

To gain further insight into the physical as well as chemical processes by which coal particles disintegrate in the early stages of dissolution, scanning electron micrographs of coals and undissolved residues can be examined. Figures 4-6(a),(b),(j) show unreacted particles of Wyodak-Anderson coal at successively increasing magnifications of 33X, 133X, and 2,000X. As has consistently been observed, the coal is fairly featureless and there is no evidence of cracks or channels. Figures 4-6(c) and (i) show West Kentucky 9,14 coal residue, after pyridine extraction, from a conversion in tetralin without stirring at 840°F for 60 seconds. A total of 87% of the organic material has been

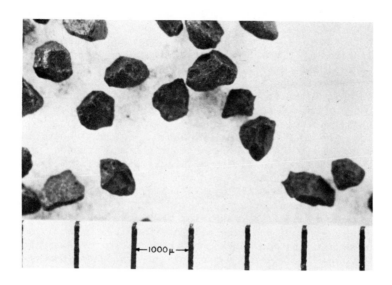

(a) Parent Coal.

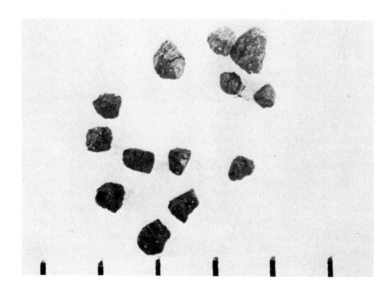

(b) After Conversion (~70%)
2 Min. 425°C
FIGURE 4-5. Wyodak Coal Particles.

removed, but the particles still retain at least their original
size; now, however, there is the appearance that the particles
consist, at least in part, of clumps of smaller domains. These
particles contain about 50% mineral matter.

Figure 4-6(d) shows four particles of the pyridine extracted
residue of Wyodak-Anderson coal that have been subjected to
conversion in tetralin at 815°F without stirring for 120 seconds;
69% of the organic material was removed. At 133X magnification,
each of the particles in Figure 4-6(d) shows strikingly unique
features. The four particles are shown at 2,000X magnification in
Figures 4-6(e)-(h); these four photographs are arranged to show
portions of the four particles of Figure 4-6(d) in the same relative positions. Figure 4-6(e) shows a portion of the upper left-
hand particle in Figure 4-6(d); it is fairly flat, but riddled
with tiny cracks and covered with fine particles. A portion of
the upper right particle, magnified in Figure 4-6(f) shows a
region where some of the original material is apparently missing.
This could be where one maceral has been dissolved away, leaving
behind another in which it was originally encased. However, the
smooth "shiny" appearance suggests the possibility that the
material we see has been melted; the open areas could be bubbles.
Figure 4-6(g) shows a region similar to the West Kentucky 9,14
residue particles shown in Figure 4-4. Figure 4-6(h) shows an area
of the lower right particle of Figure 4-6(d); this particle has
been split by massive cracks.

It appears that a number of physical processes are involved in
particle disintegration and dissolution, possibly as a function of
the macerals involved.

The information available at this time indicates that coal
particles are relatively featureless initially and will not break
up even with stirring at reaction conditions if there is no conversion to soluble form. As organic material is removed, the surface becomes eroded and a macropore system may form at the surface
and extend into the bulk. The micropore system is not drastically
changed. If the system is violently stirred, the particles break
up as dissolution progresses. In addition, dissolution and reactions do occur within the particle, further weakening it. At some
point, with stirring, the particle is broken into pieces too small
for intraparticle diffusion to be significant. It is likely that
reactions occurring within the particle and in the bulk solution
are different, contributing differently to the overall product
distribution and composition.

Information such as the above sheds light on the mode of coal
dissolution and particle disintegration and could be useful in
evaluating coal candidates for liquefaction, in developing processes, and in interpreting laboratory results.

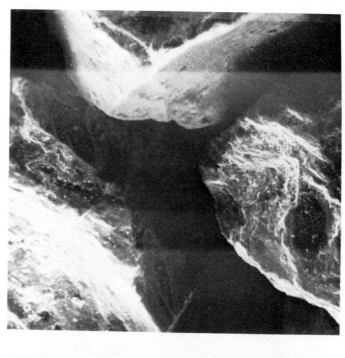

FIGURE 4-6. (a) 425-600μ particles of Wyodak Coal. Magnification 33X. (b) Same as (a). Magnification 133X.

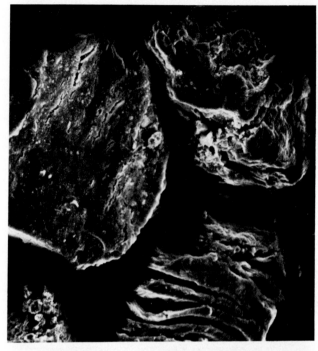

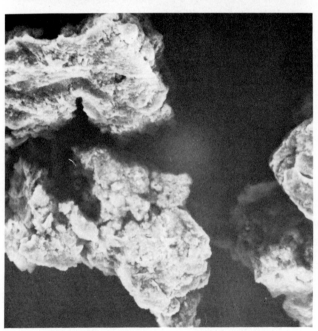

FIGURE 4-6. (c) Residue of conversion of 212-600μ West Kentucky Coal at 450°C, without stirring, in tetralin, for 60 sec.; conversion: 86.6%. Magnification 133X. (d) Residue of conversion of 425-600μ Wyodak Coal at 435°C, without stirring, in tetralin, for 120 sec.; conversion: 69%. Magnification 133X.

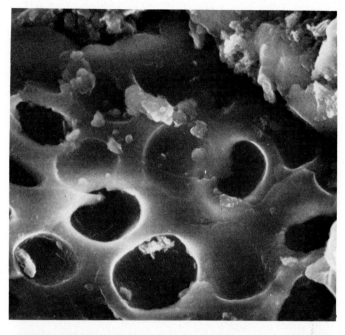

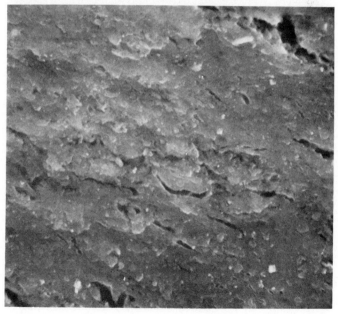

FIGURE 4-6. (e) A portion of Figure (d). Magnification 2000X. (f) A portion of Figure (d). Magnification 2000X.

FIGURE 4-6. (g) A portion of Figure (d). Magnification 2000X. (h) A portion of Figure (d). Magnification 2000X.

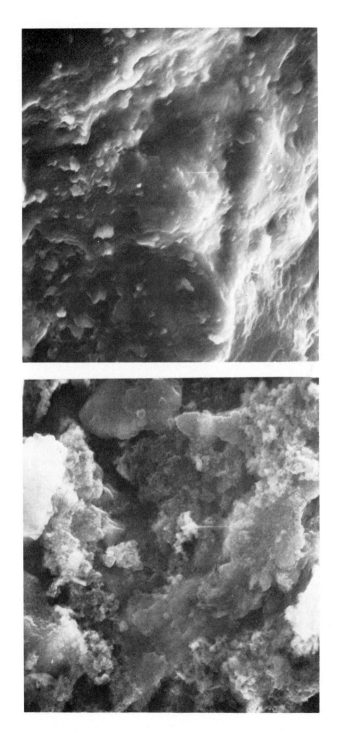

FIGURE 4-6. (i) A portion of Figure (c), residue of converted West Kentucky Coal. Magnification 2,000X. (j) A portion of Figure (a), Wyodak Coal unconverted. Magnification 2,000X.

III.   COMPOSITION VARIATIONS OF COALS AS A FUNCTION
       OF PARTICLE SIZE

     In attempting to evaluate intraparticle mass transfer limita-
tions during coal liquefactions often conversion of different
sized particles is studied. However, one complication with this
technique is that the composition of different sized particles
may be different.
     Elemental analyses and ash contents of different sizes of
some coals can be different as shown in Table 4-3 for different
particle sizes of West Kentucky 9,14 and Illinois #6 (Monterey)
coals. It can be seen that for the Kentucky coal there are only
small variations in overall elemental analysis as a function of
size; ash shows the largest variation (3.4%). The two sizes
(45-75μ and 425-600μ) are nearly identical in composition. However, there are definite trends observed in the case of the
Illinois #6 coal. As particle size increases, C, H, & O increase
and sulfur and ash decrease. The two sizes of interest show
significant differences. Most important, the smaller size contains much more ash.
     Interestingly, the breakdown of sulfur into types shows a
slightly different story. Although for West Kentucky 9,14 coal
total ash and total sulfur are nearly the same, the larger size
contains significantly less pyritic sulfur. The reverse is true
for Illinois #6 (Monterey); ash and total sulfur are very different, but pyritic sulfur is the same for the two sizes.
     These differences would clearly have to be considered in
comparisons of reactivity as a function of particle size. Such
data must also be considered in assessing the possible catalytic
activity of coal mineral matter.

IV.   PARTICLE SIZE EFFECTS AT SHORT TIMES

     Many attempts have been made to determine whether or not
there are particle size effects in coal liquefaction and the
possible significance of mass transport phenomena and diffusion
limitations.
     Published data on particle size effects were recently re-
viewed in a thesis (4-13) and showed inconsistent conclusions
in the literature. Citations of references 4-14 through 4-19
are excerpted from that thesis.
     Researchers at Spencer Chemical Company (4-14) reported that
based on their literature survey, particle size was considered to
be a variable of secondary importance.
     Ashbury (4-15) studied the benzene pressure extraction of
coal. Particle sizes of 4-8 mesh, 16-20 mesh and 60-80 mesh
coal extracted at temperatures between 428°F and 560°F gave approximately the same yield, while micron size coal gave a higher yield.

TABLE 4-3. Composition of Coals Versus Particle Size

Size, μ	Kentucky				Monterey			
	<45	45-75	75-425	425-600	<45	45-75	75-425	425-600
C	69.99	66.91	64.63	66.92[a]	64.57	63.09	69.96	69.10
H	4.31	4.43	4.27	4.42	3.74	3.89	4.30	4.20
O[b]	10.07	10.89	10.39	10.03	9.15	8.65	10.42	10.01
N	1.46	1.55	1.64	1.65	1.14	1.31	1.28	1.26
S	2.43	3.45	3.97	3.08	4.71	6.26	4.02	3.93
Ash	11.73	12.77	15.10	13.08	16.68	16.79	10.03	11.50
Sulfur[c]								
Pyritic	1.25			0.77		0.72		0.70
Sulfate	0.75			0.66		1.07		0.35
Organic	1.67			1.37		2.59		2.95
Total:	3.67			2.80		4.38		3.97

[a]CH analyses were low; these were increased to make total = 100%, using the same C/H ratio as reported in the low analyses. C and H are analyzed together in a single determination.

[b]Analyses by our Paulsboro Laboratory.

[c]By Commercial Testing and Engineering. Pyritic, Sulfate, and Total are direct; Organic is by difference.

Anderson, et al. (4-16) studied the extraction of a high volatile bituminous coal in tetralin in the temperature range of 117°F to 189°F under the influence of ultrasonic energy. He reported a more rapid initial rate of extraction with 220-270 mesh particles than with 40-60 mesh particles, but found the final yield to be the same.

Heilpern (4-17) used extraction with anthracene oil as a method of approximate study of the plastification and peptization of very large particles of different types of coals. He used a rocking autoclave operated at 756°F for 2 hours with a 3:1 ratio of anthracene oil to coal, and coal sizes of 4-3 mm, 3-2 mm, and 2-1.5 mm. The 4-3 mm and 3-2 mm showed a low solubility and the 2-1.5 mm much higher. He concluded that "particle" size was of considerable importance in the extraction of coal with anthracene oil.

Jenny (4-18) studied the liquefaction of Midlothian coal in tetralin in a hydrogen atmosphere. He used a rocking bomb autoclave. His operating conditions were 1200 psi initial hydrogen pressure, 224°F final temperature and 4 hours reaction time. He reported a decrease in percent liquefaction as particle size decreased from 100-40 mesh to -325 mesh. He reported, however, that when he simulated the rocking motion of his reactor with a beaker of tetralin and -325 mesh coal most of the coal did not become wetted, but merely floated on the surface, whereas the larger particles were more easily wetted.

In what is probably the most applicable and reliable study available, workers at Consolidation Coal Company (4-19) studied the kinetics of coal extraction with pure hydrogen donor solvents. They used a micro-autoclave that could be heated to reaction temperature in 2.5 minutes, and cooled to room temperature in 0.5 minutes. The autoclave had a volume of 30 ml and was shaken vertically at 2300 cycles per minute. They reported that their initial runs with two size fractions, i.e., 100-200 mesh and 28-48 mesh, gave identical results.

To determine the effect of the initial particle size on the course of coal liquefaction, short contact time runs were performed by MRDC and Princeton University (4-2,4-3) with Wyodak-Anderson, Illinois #6 (Burning Star), and Illinois #6 (Monterey) coals starting with different particle size ranges (45μ to 600μ). The results are presented in Table 4-4 (balances for pairs of runs that were otherwise nearly identical) and Figures 4-7 and 4-8 (plots of conversion vs. time). It can be seen that there is little or no difference in behavior of coal particles of different sizes. Thus, intraparticle mass transport limitations do not appear significant in coal liquefaction processes which utilize H-donor solvents.

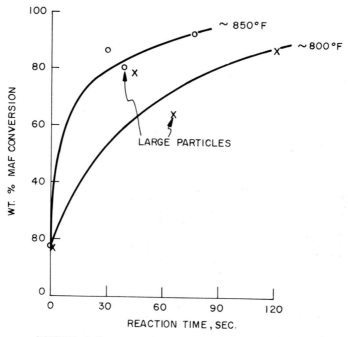

FIGURE 4-7. *Burning Star Coal Conversion*

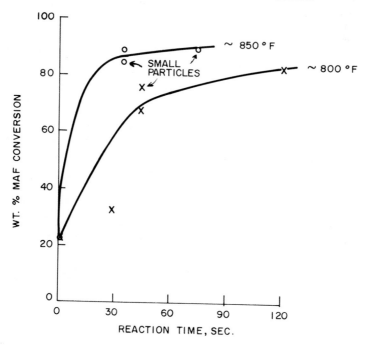

FIGURE 4-8. *Monterey Coal Conversion*

TABLE 4-4.   *Particle Size Effects*
(850-860°F, 1480-1630 psig H_2, .5-.6 min synthetic solvent/coal = 6-9)

	Illinois #6 (Burning Star)		Illinois #6 (Monterey)	
Particle Size (μ)	45-75	425-600	45-75	425-600
MAF Conversion	84.6	78.5	84.7	87.9
SRC Yield	71.9	71.0	66.6	64.1
MAF Residue Yield	15.4	21.5	15.1	20.2

The possibility was also considered that particle size might
influence the chemistry of coal conversion and thus affect the
chemical composition of the SRCs. The distributions of the SESC
fractions were compared as a function of either SRC yield or the
yield of fraction 3. No significant deviations due to particle
size effects were found.

V.   THE EFFECTS OF COAL PRETREATMENT ON LIQUEFACTION

In the storage, handling, drying, comminution, etc. before
liquefaction, coals may be subjected to significant thermal or
oxidative pretreatment. There have been studies on coal oxidation
in such areas as the effect on the macerals in lignite (4-24) or
the effect on coking properties of bituminous coal (4-25) but
there has been little work on the effects on liquefaction.
Some information has been reported (4-2) on the kinetics of
the oxidation and the TGA pyrolysis of oxidized West Kentucky 9,14
coal. Oxidation appears to occur via a fast surface reaction and
a diffusion-controlled slow reaction within the pores. Oxidation
reduces the ultimate weight loss on pyrolysis (to 1300°F in He)
only slightly but does change the weight-temperature profile.
Weight loss on pyrolysis of oxidized coal begins at a lower temperature and is more gradual. Oxidation also substantially reduces
the free swelling index of a high volatile bituminous A coal
(4-2).
Three portions of 45-75μ West Kentucky 9,14 coal were oxidized. The first was simply exposed to air at room temperature
for 6 days; TGA experiments indicate that this coal picked up
about 0.3 to 0.4 weight percent oxygen. This sample was then
exhaustively (Soxhlet) extracted with pyridine and yielded the
same 29% soluble material as an untreated sample, although the

extraction rate as judged by the color of the cycling pyridine was slower. Two additional samples were converted at 800°F in H_2 and synthetic solvent for 2 minutes: one used in run 66 was first heated in air for 17 hours at 295°F and gained 1.5% weight; another used in run 100 was heated in air for 5 days at 300°F and gained 6.0% weight. These runs are compared to a similar run with untreated coal. Elemental analyses are given in Table 4-5, and analyses are shown in Table 4-6.

The mild oxidation had little or no effect but the severe oxidation significantly reduced conversion to pyridine soluble products. This is consistent with Neavel's observation (4-21) that oxidation of a high volative C bituminous coal did not decrease conversion (to benzene solubles, in tetralin at 750°F for 30 minutes) until more than about 2% oxygen had been added. By about 4% oxygen addition the conversion was approximately halved.

The elemental analyses of the coals show that heating in air not only increases the oxygen content but rather significantly decreases the H/C mole ratio of West Kentucky 9,14 coal. The elemental analyses of the SRC's are very similar, although increasing coal oxidation decreases the oxygen content of the SRC produced. Perhaps initially present oxygen is converted to a more

TABLE 4-5.   *Conversions After Oxidative Pretreatment of West Kentucky 9,14 Coal*

(45-75 μm, 800°F, ~1500 psi H_2, 2 minutes, synthetic solvent, solvent/coal ~6)

Run Number	66	67	100
Treatment	Oxid.		Oxid.
MAF Conversion, wt %	79.10	80.60	55.70
Yield of SRC	47.90	68.94	52.65
SESC Fraction (% of SRC)			
1 HEXANE	2.18	1.40	1.65
2 HEX/BZ	5.98	6.97	7.19
3 CLF	4.71	5.55	7.73
4 CLF/ET	13.31	7.43	15.58
5 ET/EOH	19.72	13.14	12.80
6 MEOH	12.91	19.22	14.94
7 CLF/OH	5.50	14.79	9.85
8 THF	10.64	7.62	5.29
9 PY/OH	25.06	23.87	24.96

TABLE 4-6. Elemental Analyses for Oxidative Pretreatment of West Kentucky 9,14 Coal

	Coal	Coal[a] Oxid.	Coal[b] Oxid.	66 SRC	67 SRC	100 SRC	66 Residue	67 Residue	100 Residue
Run Number				66	67	100	66	67	100
Sample	Coal	Coal[a]	Coal[b]	SRC	SRC	SRC	Residue	Residue	Residue
Treatment	O	Oxid.	Oxid.						
Fraction	O								
Percent of Coal	100.00	100.00	.00	47.90	68.94	.00	25.00	.00	.00
Percent of SRC	.00	.00	.00	100.00	100.00	96.00	.00	.00	.00
C	73.06	70.95	66.07	81.10	82.30	80.46	62.09	55.72	70.82
H	5.00	4.60	3.73	5.80	5.73	5.55	3.15	3.03	3.53
O	9.17	12.14	16.58	7.56	6.78	4.99	6.45	3.91	8.09
N	1.47	1.25	1.54	1.74	1.80	1.59	1.44	1.18	1.97
S	2.97	3.22	2.65	1.25	2.12	2.27	3.77	4.46	2.04
Ash	8.33	7.84	9.43	2.55	1.27	5.82	23.10	31.70	13.55
H/C	.82	.78	.68	.86	.84	.83	.61	.65	.60

[a] Used in run 66.

[b] Used in run 100.

easily removed form. Aside from those due to varying ash content,
the only significant differences in the residue analyses are that
residues from oxidized coals contain more oxygen.

The SESC analyses show that prior oxidation of coal results in
an SRC that contains more of fraction SESC-4 (phenols) and less of
fractions SESC-6 (high heteroatom) and SESC-7 (polar polyphenols).
It is interesting to note that there is no increase in ethers
(SESC-3) or highly polar polyphenolic materials (SESC-8, 9). The
decrease in SESC-7 is particularly surprising.

Runs were also performed to examine the effect of thermal
pretreatment alone, partly to uncouple the oxidation results from
the heating occurring in that pretreatment. A sample of West
Kentucky 9,14 coal was heated for 20 hours at 480°F in He (initial
pressure 200 psi at 70°F). Weight loss was 2.8%. TGA pyrolysis
of this thermally pretreated coal resulted in a weight loss of 21%
up to 1110°F in He; the untreated parent coal lost 29% weight.

The results of liquefaction of the thermally pretreated West
Kentucky 9,14 coal are given in Tables 4-7 and 4-8 and compared

TABLE 4-7.   *Conversions After Thermal Pretreatment*
of West Kentucky 9,14 Coal

($800°F$, 1500 psi H_2, 2 minutes, synthetic solvent)

Run Number	43	67
Treatment	Heat[a]	
Solvent/Coal	13.00	7.40
MAF Conversion, wt %	69.90	80.60
CO_2	1.26	.55
Yield of SRC	44.70	68.94
SESC Fraction (% of SRC)		
1 HEXANE	.85	1.40
2 HEX/BZ	6.86	6.97
3 CLF	12.43	5.55
4 CLF/ET	17.91	7.43
5 ET/EOH	10.33	13.14
6 MEOH	12.55	19.22
7 CLF/OH	3.19	14.79
8 THF	9.55	7.62
9 PY/OH	26.33	23.87

[a]*Twenty hours at 480°F in He.*

TABLE 4-8. *Elemental Analyses for Thermal Pretreatment*
of West Kentucky 9,14 Coal

Run Number			43	67	43	67
Sample	Coal	Coal	SRC	SRC	Residue	Residue
Treatment		Heat[a]				
Percent of Coal	100.00	98.00	47.71	68.94	35.94	27.80
C	73.06	72.66	81.68	82.30	65.50	55.72
H	5.00	4.85	5.76	5.73	3.62	3.03
O	9.17	10.03	8.49	6.78	12.45	3.91
N	1.47	1.40	1.63	1.80	1.86	1.18
S	2.97	3.40	-	2.12		4.46
Ash	8.33	7.60	.84	1.27	16.57	31.70
H/C	.82	.80	.85	.84	.66	.65

[a]*Twenty hours at 480°F in He.*

to a similar conversion of untreated West Kentucky 9,14 coal. In
Table 4-7 it can be seen that the thermal treatment reduced conversion by about 10% under the specified conditions. The selectivity to SRC is about the same in the two cases, but thermal
treatment increases the yield of CO_2.

From the elemental analyses in Table 4-8 we see that the
thermal pretreatment has little effect on the coal, SRC, or
residue except for an increased oxygen content of the SRC and
residue. The SESC distributions show that the SRC from the
treated coal tends to contain less polar material and, like the
oxidized West Kentucky 9,14, there is more SESC-4 and less SESC-6
and 7.

Heat treated samples of Wyodak-Anderson coal were also studied.
A sample of Wyodak-Anderson coal was heated in He for 17 hours at
390°F and then for 17 hours at 490°F; weight loss was 2.2%. Water
(and H_2S from the West Kentucky 9,14 coal) was detected by mass
spectroscopic analysis of the helium after these experiments.
Two additional Wyodak-Anderson samples were prepared; one heated
for 17 hours at 480°F in He, and one heated for 6 days at 480°F
in argon. The TGA pyrolysis of Wyodak-Anderson coal heated for
17 hours at 480°F is shown in Figure 4-9 and compared to untreated Wyodak-Anderson. Thermal pretreatment alone brought about
less weight loss, but the loss occurred with about the same weight-
temperature profile as for the untreated coal.

Soxhlet extraction of the Wyodak-Anderson heat-treated coal
(17 hours, 480°F) yielded 22% pyridine soluble material; the
parent coal yields 11.5%. The N_2 surface area of the treated

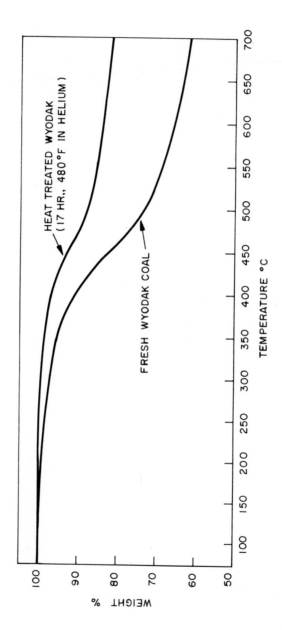

FIGURE 4- 9 . TGA Thermograms; Pyrolytic Yields of Fresh and Heat Treated Wyodak Coal.

Wyodak-Anderson coal was 1.2 m^2/g; for the untreated coal it was 2.5 m^2/g.

Table 4-9 shows the elemental analysis of the coals before and after thermal treatment and Table 4-10 shows the effect of this treatment on liquefaction behavior. In the absence of hydrogen gas, conversion of the mildly heated coal was lower than that of unheated coal. However, in the presence of hydrogen gas, little effect on conversion or product distribution was noted. Some change in the H/C ratio of the SRC was noted in that the SRC from the heat treated coal had a lower hydrogen content than the SRC from untreated coal. This probably reflects the fact that heating the coal lowered its H/C ratio. The residue analyses of the two sets of conversions were similar.

Summary

Summarizing the effects of particle size one concludes that although there are some variations in conversion when different particle size ranges are used, there is clearly no consistent pattern that would suggest any significant mass transport limitations. This is also consistent with the observations that surface areas and pore volumes (less than 300 Å radius) are very low and, under the conditions in which these runs were made, particles disintegrate very rapidly to sub-micron size. It is most likely that if there are any real particle size effects at all, they are due to variations in mineral and maceral contents of the size ranges or minor variations in the conversion procedure.

Similarities of Thermal and Oxidative Pretreatments

- Both reduce pyrolytic weight loss of coals; the thermal treatment has a greater effect but does not change the weight-temperature profile as oxidation does.

- Both can reduce liquefaction conversion and result in more light gas formation. The effects are greater for severe oxidation. Bituminous coal is more sensitive to thermal treatment than sub-bituminous coal.

- Both make a less polar SRC, especially increasing fraction SESC-4 (phenols) and decreasing SESC-6 and SESC-7.

- Both result in a higher residue oxygen content.

TABLE 4-9. Coal Analysis After Heat Treatment of
Wyodak-Anderson Coal

Treatment	None	Heat[a]	None	Heat[b]
C	65.94	66.51	71.82	70.90
H	5.36	5.32	5.20	4.28
O	18.10	16.65	17.12	21.32
N	1.00	1.02	0.90	1.10
S			0.30	
Ash	9.60	10.50	4.66	2.40
H/C	0.98	.96	.87	.72

[a] 17 Hours at 390°F then 17 hours at 480°F in He.

[b] Six days at 480°F in He.

TABLE 4-10. Conversions After Heat Treatment of
Wyodak-Anderson Coal

(synthetic solvent, solv/coal = ∿10, 2 minutes)

Treatment	None	Heat[a]	None	Heat[b]
Temperature, °F	800	800	814	800
Atmosphere	Nitrogen	Nitrogen	Hydrogen	Hydrogen
H_2 Pressure	0	0	1400	1380
MAF Conversion	54.8	49.3	60.3	62.2
SRC Yield	37.8		52.3	56.4
SRC Composition (%)[c]				
Oils	8.4		14.2	13.7
Asphaltenes	26.9		41.3	38.4
Asphaltols	64.4		44.5	47.9
SRC Elemental Analysis (% MAF)				
C	82.91		81.15	80.68
H	6.44		7.02	5.83
O	10.42		10.26	12.29
N	1.16		1.18	1.20
S	−		.93	−
H/C	.93		1.04	.87

[a] 17 Hours at 390°F then 17 hours at 480°F in He.
[b] Six days at 480°F in He.
[c] As estimated from SESC analyses.

Differences of Pretreatments

- Mild oxidation does not change Soxhlet pyridine solubility. Thermal treatment increases solubility (Wyodak-Anderson).

- Oxidation increases oxygen in coal and decreases H/C mole ratio; thermal pretreatment does not grossly change elemental analysis of coal.

- Oxidation decreases oxygen content of SRC; thermal treatment increases oxygen in SRC.

- Thermal or oxidative pretreatment of coal does not appear to be beneficial to the SRC process, although it is noteworthy that mild oxidation is not very detrimental. This indicates that extreme care to prevent any oxidation of feed coals may not be necessary in commercial operations.

REFERENCES

4- 1. H. Gan, S. P. Nandi, and P. L. Walker, Fuel, 51, 272 (1972).

4- 2. D. D. Whitehurst, M. Farcasiu and T. O. Mitchell, "The Nature and Origin of Asphaltenes in Processed Coals", EPRI Report AF-252, First Annual Report Under Project RP-410, February 1976.

4- 3. T. O. Mitchell and D. D. Whitehurst, "Short Contact Time Coal Liquefaction", ACS Division of Fuel Chemistry, Preprints page 127, San Francisco, August 1976.

4- 4. E. G. Plett, A. C. Alkidas, F. E. Rogers and M. Summerfield, Fuel, 56, 241, (1977).

4- 5. M. Monta and K. Hirosawa, Nenryo Kyokai-shi 53 (564), 263 (1974).

4- 6. P. L. Walker, W. Spackman, P. H. Given, E. W. White, A. Davis and R. G. Jenkins, "Characterization of Mineral Matter in Coals and Coal Liquefaction Residues", EPRI Report AF-417, June 1977.

4- 7. R. G. Jenkins, "Mechanisms Involved in the Formation of Reactor Solids", Section 7 of Proceedings of EPRI Contractor's Conference on Coal Liquefaction, Palo Alto, CA., May 1978.

4- 8. W. Francis, "Coal - Its Formation and Composition", Edward Arnold Ltd., London, 1961, p. 710.

4- 9. I. G. C. Dryden, Fuel, 20, 145 (1951).

4-10. "Solvent Refined Coal Studies", Technical Report Prepared for National Science Foundation Covering the Period May 31, 1974 to May 31, 1975, on NSF Grant 38701, Auburn University Chemical Engineering Department, Auburn, Alabama.

4-11. J. A. Guin, A. R. Tarrer, Z. L. Taylor and S. C. Green, "A Photomicrographic Study of Coal Dissolution", American Chemical Society 169th National Meeting, Division of Fuel Chemistry, Preprint, Vol. 20, No. 1, pp. 66-76, April 6-11, 1975.

4-12. A. R. Tarrer, Personal Communication; Results Presented at AIChE Meeting in Los Angeles, California, November 1975.

4-13. S. C. Greene, Jr., "Investigation of the Mechanism of Dissolution of Coal: Effect of External Mass Transfer and of Intraparticle Diffusion", Master's Thesis, Auburn University, December 11, 1974.

4-14. D. L. Kloepper, T. F. Rogers, C. H. Wright and C. W. Bull, Research and Developement Report No. 9, Solvent Processing of Coal to Produce a Deashed Product, Office of Coal Research, Department of the Interior, Washington, D. C., 1965.

4-15.   R. S. Ashbury, "Action of Solvents on Coal", <u>Industrial and Engineering Chemistry,</u> <u>26</u>, 1934, pp. 1301-1304.

4-16.   Larry L. Anderson, M. Yacob Shifai and George R. Hill, "Ultrasonic Energy Effects on Coal Extraction by a Hydrogen Donor Solvent", American Chemical Society, Division of Fuel Chemistry, Reprints 14 (1), 1970, pp. 115-143.

4-17.   Heilpern, Stanislawa, "Investigation of the Mechanism of Phenomena Occurring in the Plastic State of Bituminous Coal Treated by Solvent Extraction", <u>Koks-Smola-Gas,</u> <u>2,</u> 1957, pp. 8-11.

4-18.   Jenny, Max Frederick, "The High Pressure Hydrogenation of Midlothian Coal, Part II", Unpublished M. S. Thesis, Blacksburg, Virginia, 1949.

4-19.   George P. Curran, Robert T. Struck, and Everett Gorin, "The Mechanism of the Hydrogen Transfer Process to Coal and Coal Extract", American Chemical Society, Division of Fuel Chemistry, Preprints, 10 (2) 1966, pp. 130-148.

4-20.   G. P. Curran, R. T. Struck and E. Gorin, <u>I and EC Process</u> Design and Development, 6, pp. 166-173 (1967).

4-21.   R. C. Neavel, Fuel, <u>55</u>, 237 (1976).

4-22.   P. H. Given, D. C. Cronauer, W. Spackman, H. L. Lovell, A. Davis, and B. Biswas, Fuel, <u>54</u>, 34 (1975).

4-23.   Ibid, Fuel, <u>54</u>, 40 (1975).

4-24.   S. P. Valceva, K. L. Markova, D. D. Rouschev and E. E. Bekyarova, Fuel, <u>55</u>, 173 (1976).

4-25.   A. Y. Kam, A. N. Hixson, and D. D. Perlmutter, <u>Ind. Eng.</u> Chem. Proc. Des. Dev. <u>15</u>, 416 (1976).

4-26.   R. G. Jenkins, The Pennsylvania State University, Personal Communication.

4-27.   M. M. Dubinin and G. M. Plavnik, <u>Carbon</u> <u>2</u>, 261 (1961); Carbon <u>6</u>, 183 (1968).

4-28.   D. D. Whitehurst, M. Farcasiu, T. O. Mitchell, J. J. Dickert, Jr., "The Nature and Origin of Asphaltenes in Processed Coals", EPRI AF-480, Research Project 410-1, Second Annual Report, March 1976-February 1977.

Chapter 5

COAL RANK AND LIQUEFACTION

I. INTRODUCTION

Various behavioral differences have been noted between subbituminous coals and bituminous coals, particularly in dissolution rates, product yields of different chemistry, and sensitivity to side reactions (5-1, 5-2). In this chapter, how coal
rank or composition affects coal liquefaction is discussed (5-8).

Generally rank is determined by measuring the heat content or
calorific value of a coal and can be used as a measure of its
degree of coalification (increasing rank means more coalified).
The heat content is a function of the elemental composition of
the organic portion of the coal and its moisture and ash content.
Since most coals suitable for liquefaction contain 5-6% H and
~1-2% N, the heat content (or rank) increases with increasing
C/(O+S) ratio. Thus as a crude approximation, the rank of a coal
can be represented by its carbon content on a moisture and ash
free basis.

Many scientists have attempted to correlate the extent of
conversion to benzene-soluble products at long times with coal
rank. Much of this effort, however, did not consider the
benzene-insoluble/pyridine-soluble products and, therefore, cannot be used as a guide to the total conversion of coal to another
form. Some workers (5-3, 5-4) did report the acetone-soluble
products, which is more instructive.

Storch and co-workers (5-4) evaluated a number of coals and
concluded that two compositional parameters must be considered in
coal liquefaction - the petrography and the MAF carbon content of
the coal. Certain macerals were less reactive than others, and
for a given maceral type, a minimum in conversion occurred with

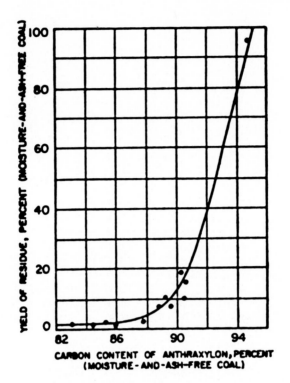

FIGURE 5-1.   Yield of Residue from Reactive Macerals (5-3).

increasing carbon content.   The petrography used at that time is
different from present assignments, but the results can be used
as a guide.   Figure 5-1 shows the yield of residue from a reactive
maceral (anthraxylon) as a function of MAF carbon.   Above ∿89% C,
even reactive macerals produce substantial residue yields and thus
poor conversion.

Workers at the Pennsylvania State University (Penn State)
(5-5) and Gulf Research and Development have recently concluded,
based on a series of high-vitrinite coals, that the yield of
benzene-soluble product is very low for coals having greater than
90% MAF carbon.   Maximum conversions to benzene-soluble products
were observed in the range of 82-84% MAF carbon.   The conversion
or oil/asphaltene ratio correlated well with the content of reactive macerals and the degree of coal conversion, but poorly
with the geological origin of the coal.

The Institute of Technical Chemistry in Germany has compared
the relative ease of coal liquefaction to benzene-soluble product

(no catalyst or organic carrier solvent was used) (5-6). Their
results indicate that in the range from 52% C (dry leaves and
peat) to 84% C, conversion tends to increase with increasing H/C
or O/C ratios of the feed. Carbon monoxide tends to give higher
conversions at all ranks than does hydrogen gas.

MRDC and Penn State examined a series of high vitrinite coals
with reactive maceral contents of 82-91% and varying rank; the
DMMF carbon contents ranged from 55-90%. In one set of experiments, maceral variations are minimal. Thus, rank should be
responsible for any behavioral performance differences. These
five coals are described in Tables 5-1 and 5-2, along with
additional coals which were not reacted but were used for
aromatic carbon content correlations.

Two series of experiments were performed using these coals.
The experimental conditions were for 3 or 90 minutes, at 803°F,
under H_2, in a solvent (SS*8.5) consisting of 2%γ-picoline, 18%
p-cresol, 8.5% tetralin, and 71.5% 2-methylnaphthalene. The conditions were selected such that conversion at 3 minutes would be
incomplete, allowing the Penn State analyses to indicate maceral

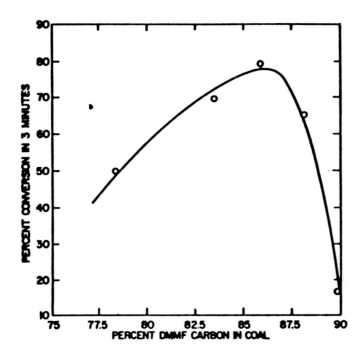

FIGURE 5-2. Conversion vs. Percent DMMF Carbon.

TABLE 5-1.   Analyses of Coals

		PSOC-106	PSOC-235	PSOC-256	PSOC-312
Mine Location	Name of Coal	PSOC-106	PSOC-235	PSOC-256	PSOC-312
	State	Indiana	Colorado	Pennsylvania	Arizona
	County	--	--	Clearfield	--
	Seam	Indiana #1 Block	Colorado E--E2 Bed	Lower Freeport	Red
	Mine	--	--		--
Proximate Analysis*	% Moisture (as rec.)	6.40	3.87	0.75	10.70
	% Ash (as rec.)	13.46	7.35	6.69	5.90
	% Volatile Matter	34.17	41.09	25.98	43.15
	% Fixed Carbon	51.45	51.26	67.27	50.24
	BTU/lb. (as rec.)	11,549	12,739	14,489	11,378
	BTU/lb.	12,339	13,252	14,598	12,741
	Free Swelling Index	0.5	--	8.5	0.5
Ultimate Analysis*	% C	70.85	74.07	81.77	72.58
	% H	4.72	5.15	4.57	5.38
	% O**	8.99	10.96	4.58	13.76
	% N	0.47	1.49	1.63	1.30
	% S (total)	0.59	0.68	1.75	0.37
	% S (pyritic)	0.13	0.09	1.12	0.07
	% S (organic)	0.46	0.58	0.62	0.29
	% S (sulfate)	0.0	0.01	0.01	0.01
	% Cl	0.0	0.02		0.05
	% Ash	14.38	7.65	4.58	6.61

	PSOC-330	PSOC-331	PSOC-372	PSOC-405
Name of Coal				
State	Pennsylvania	Pennsylvania	Kentucky	Oklahoma
County	--	--	--	LeFlore
Seam	Middle Kittanning	Brookville	Imboden	Lower Hartshorne
Mine	--	--	--	--
Proximate Analysis*				
% Moisture	6.14	5.31	1.04	4.36
% Ash (as rec.)	5.73	4.78	5.47	7.87
% Volatile Matter	37.93	37.86	34.67	18.03
% Fixed Carbon	55.96	57.09	59.80	73.74
BTU/lb. (as rec.)	12,887	13,413	14,410	13,602
BTU/lb.	13,730	14,165	14,561	14,222
Free Swelling Index	5.0	4.5	8.0	--
Ultimate Analysis*				
% C	76.93	79.73	80.41	81.49
% H	5.09	5.33	5.22	4.29
% O**	7.39	7.57	6.32	3.77
% N	1.71	1.54	1.24	1.55
% S (total)	2.77	0.78	1.28	0.67
% S (pyritic)	2.08	0.06	0.48	0.06
% S (organic)	0.67	0.71	0.80	0.61
% S (sulfate)	0.02	0.0	0.0	0.0
% Cl	0.16	0.15	0.02	0.14
% Ash	6.11	5.05	5.53	8.23

Mine Location

*All analyses are given on a dry weight basis, unless otherwise stated.
**By difference.

127

TABLE 5-2. Maceral Composition* of High Vitrinite Coals

PSOC	R_O	VIT	PVIT	FUS	SFUS	MAC	MIC	SPOR	RES	CUT	% Ash**
312	0.44	80.2	5.7	4.6	4.9	1.0	1.2	2.2	0.2	0.0	5.9
330	0.76	78.3	5.9	1.5	2.8	0.5	7.6	3.1	0.3	0.0	5.7
372	1.00	71.4	13.9	3.0	1.2	0.5	3.8	5.6	0.6	0.1	5.5
256	1.26	67.0	15.2	3.8	3.9	0.5	9.6	0.0	0.0	0.0	6.7
405	1.68	75.0	6.2	7.3	7.3	0.2	4.0	0.0	0.0	0.0	7.9

*Macerals reported as volume percentage on DMMF basis

R_O – mean maximum reflectance of vitrinite (+ pseudovitrinite) in oil, %

** Data from proximate analysis of coal as received

VIT – vitrinite   MIC – micrinite
PVIT – pseudovitrinite   SPOR – sporinite
FUS – fusinite   RES – resinite
SFUS – semifusinite   CUT – cutinite
MAC – macrinite

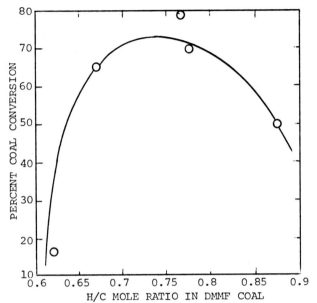

FIGURE 5-3.   Conversion vs. H/C Ratio in DMMF Coal

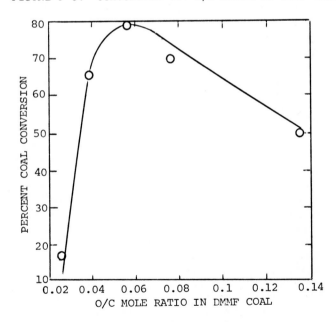

FIGURE 5-4.   Conversion vs. O/C Ratio in DMMF Coal

reactivity as a function of rank. All coal conversions will be
defined as conversion to pyridine-solubles unless otherwise
stated. These results are discussed in the next two sections.

II. INTRINSIC REACTIVITY (From Short Time Conversions)

Let us consider the series of 3 minute runs first; only the
high vitrinite series will be included in the initial discussions
in order to minimize the effects of maceral variations. Under
essentially the same conditions, conversions varied from 25% to
80%.

Some relationships between conversion (3 min., 800°F) and feed
coal composition are shown in Figures 5-2 thru 5-4. The conversions vs. DMMF carbon, H/C or O/C mole ratios all exhibit
maxima. These data are in general agreement with previously
reported work at long times (5-3, 5-4, 5-5), but in the present
case, the observed maximum in conversion occurred at 85% C instead
of at 82% C. The observed trends in conversions vs. H/C or O/C
mole ratios differed greatly from the previous report of Oelert
(5-6), but no solvent was used in his work and benzene-solubility
instead of pyridine-solubility was the conversion criterion.

Because all of the coal is not amenable to liquefaction (5-5);
one must suspect that the MAF C content should not be a good
measure of coal reactivity, and perhaps the conversion of the
"reactive" coal constituents vs. the composition of the reactive
portion of the coal would be more appropriate.

We have attempted to calculate the composition of the
"reactive" portions of the various coals using correlations
developed by others (5-3, 5-4, 5-7, 5-8, 5-9) and compare these
compositions to coal conversion. These correlations, however,
showed no improvement over the uncorrected values (5-27).

Another parameter that relates the rank of coals is the mean
maximum vitrinite reflectance (5-12). Davis et al. predicted
that the best coals for liquefaction would have mean maximum
vitrinite reflectance between 0.5 and 1.0%. In Figure 5-5, coal
conversions are shown using this rank parameter; these curves
are similar to those using % MAF carbon. 0.5% appears to be too
high as a minimum reflectance for suitable coals; Wyodak-Anderson
(Belle Ayr) has a reflectance of only 0.35%. PSOC-256 has a
reflectance of 1.33% and appears to be a reactive coal.(But 1.0%
may be too low for the upper limit).

In contrast to reports (5-3, 5-4) of long time coal conversions which show essentially equal conversions of coals having
less than 86% carbon (MAF), the present data indicate low conversions for low ranks at 3 minutes. This difference can be best
explained by a lower inherent rate of dissolution of these coals.

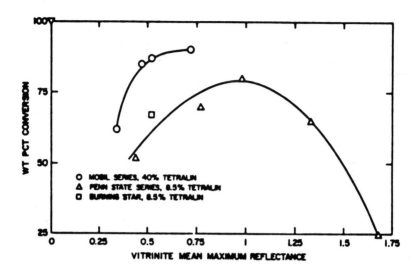

FIGURE 5-5.   Conversion vs. Reflectance

Figure 5-6(a) shows the CO_2 yield vs. rank (% MAF C in feed coal).   There is a smooth decrease in CO_2 yield with increasing rank and the lowest rank coal gave the greatest yield.   The observed CO and CO_2 yields for these short-time runs can be compared with the values predicted from past data of the Bureau of Mines (BuMines) in long-contact-time coal conversions (5-3).

YIELDS IN g/100 g COAL

CO Observed	CO Predicted	CO_2 Observed	CO_2 Predicted
0.00	0.13	0.14	0.06
0.09	0.17	0.26	0.09
0.00	0.26	0.31	0.28
0.08	0.33	0.51	0.53
0.28	0.76	1.47	2.88

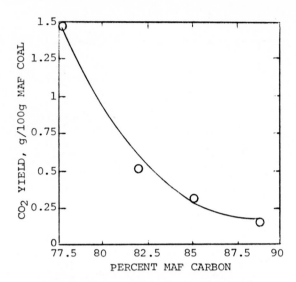

5-6(a). *CO$_2$ Yield vs. % MAF Carbon*

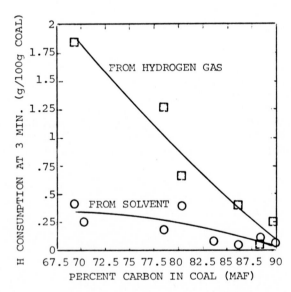

5-6(b). *Hydrogen Consumption vs. % MAF Carbon*

FIGURE 5-6. *Liquefaction Behavior as a Function of Rank*

The "predicted" values are from correlations developed (5-9) from ultimate analysis and BuMines conversion data under more severe conditions. The CO yields are lower in these short time conversions than the BuMines data. This suggests that at 3 minutes the majority of the CO production is not fully developed, (although there is a trend toward increasing CO with decreasing rank). For CO_2 we observed the predicted yields except for the lowest rank case where the eventual CO_2 yield would be very high. CO_2 is clearly produced very early in coal conversions. We always obtained less CO than CO_2 although with further conversion the CO would probably exceed CO_2 for high ranked coals. The fact that we did not observe sufficient CO_2 for the slowly dissolving low rank coal in our 3 minute run suggests that CO_2 evolution parallels coal dissolution.

Figure 5-6(b) shows the total H-consumption (from both gas and solvent) vs. rank; most of the coals consumed very little hydrogen except the lowest rank which consumed far more. H-consumption from the solvent was very low for the entire series. The high consumption by PSOC-312 was substantially from the gas; the value is surprisingly high for such a short time and might be partly experimental error, but it is consistent with the accepted hypothesis of increased H-consumption by low rank coals. For the initial dissolution (3 minutes) no correlation was observed between the H-consumption and the methane or other light gases produced.

The CH_4 yield at short time shows a maximum close to where conversion and SRC yield peak. H_2S, H_2O and CO yields in these runs were all very low; no light liquid products were obtained.

The BuMines (5-3) work indicates that at maximum conversion, the light gas yields decrease with increasing rank over the rank range in the Penn State series. In the MRDC 3 minute runs, however, the low rank coals, because they convert to pyridine-solubles more slowly, show lower light gas yields. The situation is analogous to the CO_2 yield data discussed above.

For the same reason we observed a maximum in SRC yield. At long times Storch (5-3) saw no maximum in acetone-soluble yield but Given et al. (5-6) observed a maximum (at 82% MAF C in feed coal) in the yield of benzene-soluble products. The highly-functional low-rank coals probably gave functionalized products insoluble in benzene, so that this observed maximum was due to the chemical nature of the product. This appears to be borne out by SESC analysis.

The overall picture is that these MRDC/Penn State studies, although conducted for short times, agree with prior investigations conducted at long times, except in the low rank range; pyridine-solubles formation is apparently slow enough to affect

the results. It does appear, however, that the range of coal
ranks suitable for liquefaction is somewhat greater than anticipated (5-12).

The residues from the series of five 3 minute runs with high
vitrinite coals have been optically examined in detail and compared to the parent coals (5-13(a)). Maceral analyses for the
coals and residues are given in Table 5-1 and 5-3 respectively.
A comparison of rank, conversion, and reflectance is presented in
Table 5-4.

Because of the short reaction time, differences between the
petrographies of residues and starting coals closely represent
different conversion rates for different macerals; the percentage
of vitrinite (including pseudovitrinite), which could affect data
interpretations, varied only from 81 to 86% in these coals.

The residue from the hvCb coal PSOC-312 showed the least
plasticity and agglomeration, and included more granular residue
and unreacted vitrinite, and less vitroplast, than residues from
higher rank coals. Much of the vitroplast was in fact partially
reacted vitrinite (although this category is not used in residue
analyses, to avoid confusion). Conversion was only 51% for this
coal, the second lowest of the five (not surprisingly the low
volatile coal PSOC-405 underwent the lowest conversion in Table
5-4). Highest conversion (79%) occurred for the hvAb coal PSOC-
372. Any enhancement of conversion by mineral matter was minor,
because this coal produces the lowest percentage of ash in the
group (Table 5-2). The coal with the highest ash content in fact
gave the lowest conversion (the lvb coal), showing that the
importance of rank overrides that of mineral catalysis for this
series of experiments.

The following general conclusions were drawn from examination
of these residues (5-13(a)) and consideration of the runs in which
they were products (5-8).

1. Each residue includes a higher percentage of inert
macerals than its corresponding coal, the highest percentages of
fusinite plus semifusinite being in PSOC-405 (hvAb) and its
residue. The percentage of these components in the residue from
the hvAb coal (highest conversion) is disproportionately large.

2. There is no relationship between percentage of ash in the
feed coal and percentage of granular residue produced.

3. The coal with the highest percentage of exinite (PSOC-372)
experienced the highest conversion. However, the residue from
that coal included much sporinite and apparently rank is the
overriding factor here.

4. The residues from coals with the largest amount of
micrinite included very little granular residue. This maceral may

TABLE 5-3. Residue Analyses*, High Vitrinite

PSOC	Gran Res	Semi-coke	Ceno-sph	Vitro-plast	Unreac Vit	Fus	Semi-fus	Pyrrh	Clay	Carb	Qtz	Other
312	7.9	-	-	70.0	4.5	6.8	8.1	tr.	1.5	0.4	0.7	(a) (b) (c)
330	2.6	tr.	0.6	73.5	-	3.0	9.3	5.6	4.9	-	-	(c)
372	7.4	0.5	1.9	72.4	0.6	7.8	5.3	0.8	1.5	-	tr.	(c)
256	0.6	tr.	-	82.2	-	8.6	4.7	1.7	0.9	-	0.9	(d)
405	-	2.1	-	75.6	-	8.0	9.2	tr.	0.6	3.9	tr.	-

*Values given in volume %
tr. = trace, <0.5%
(a) Ca-sulfate = tr.
(b) other inertinite = tr.
(c) sporinite = tr.
(d) sporinite = 1.5%

Gran Res - granular residue
Cenosph - cenospheres
Unreac Vit - unreacted vitrinite
Fus - fusinite
Semifus - semifusinite
Carb - carbonate minerals
Qtz - quartz
Pyrrh - pyrrhotite

TABLE 5-4. Comparison Among Rank, Conversion and Reflectance of High Vitrinite Coals and Residues

PSOC	Rank	% Conversion	Reflectance* of Vitrinite in Coal	Reflectance* of Vitroplast in Residue	% Change in Reflectance
312	hvCb	51	0.44	0.97	+120
330	hvBb	70	0.76	1.36	+ 79
372	hvAb	79	1.00	1.23	+ 23
256	mvb	65	1.26	1.46	+ 16
405	lvb	17	1.68	1.64	- 2

*Percentage of incident light, in oil

not contribute noticeably to the granular residue component, in
which case it is either reacted (unlikely) or dispersed throughout the plastic phase.

   5. Minerals are largely unchanged, except that clays are
somewhat dispersed, and pyrite has been partially reduced to
pyrrhotites. Size appears to be a factor limiting these changes.
Small (5 to 10 μ) dispersed pyrite crystals are completely converted. Larger areas retain pyrite cores. Smaller clay masses
appear preferentially dispersed.

   These observations are applicable to this set of coals only,
because they comprise a rank series, and are not meant to be
generally applicable to all coals.

   More striking trends in residue composition and characteristics are controlled by rank. The following is a partial list
of these trends (5-13(a)):

   1. Essentially no granular residue was formed by coals with
higher rank than hvAb (85.9% MAF C).

   2. Unreacted vitrinite was an important constituent only in
the residue from the hvCb coal (78.4% MAF C).

   3. The vitroplast content appeared to be largely unrelated to
rank or conversion, for coals above hvCb rank. Of course, the
absolute amounts would be expected to vary inversely with conversion.

   4. Anisotropic semi-coke formed only from the coals above
hvBb rank (83.5% MAF C). It was widespread only in the residue
from the lvb coal.

   5. The most evidence of plasticity and agglomeration was
seen in residues from hvBb through mvb ranks (88.2% DMMF C).
These residues included "frothy" particles (more pore volume than
cell wall) and cenospheres. Reaction conditions are optimal for
their plasticity in the range of temperature involved.

   6. Conversion increased with rank to hvAb, then decreased,
especially above hvb.

   7. Semifusinite in the residues from the hvCb and hvBb coals
appeared partly reacted, having rounded margins, fractures, and
preferential removal of some areas. It did not appear reacted in
the residues from hvAb rank or above, even though the hvAb coal
experienced the highest conversion. This suggests that semifusinite may be reactive only if the rank is lower than hvAb under
these conditions.

   8. Reflectance measurements of vitroplast show that for
coals of hvCb through mvb rank, the reflectance of vitroplast is
considerably higher than that of the original vitrinite. The
percentage of change in reflectance from vitrinite to vitroplast
is inversely related to rank, going from 20% at hvCb to -2%
(essentially unchanged) for the lvb (89.8% MAF C) coal and residue

(Table 5-4). This trend is so consistent that it results in the
vitroplast from hvAb vitrinite having a higher reflectance than
that from the hvBb vitrinite. These results seem to indicate
that molecular mobility (necessary for the increased molecular
alignment to increase reflectance) is greater with decreasing
rank.

The greater mobility with lower rank should enhance formation
of semi-coke. It has been reported (5-13(b)) that with reactor
solids from bituminous and subbituminous coals much mesophase
semi-coke is produced from a lower rank coal (Wyodak-Anderson).
The short residence time in this series was probably the limiting
factor, thus semi-coke was formed only where molecular alignment
was already advanced (i.e., higher rank coals). There have been
other reports that lower rank coals are slower to dissolve. Given
et al. (5-14) stated that at low conversion levels (to benzene
soluble products) in a continuous liquefaction unit, solvation of
low rank coals increases more rapidly than that of higher rank
coals with increasing residence time.

The most reactive high volatile bituminous coals are those
that develop maximum plasticity, which depends upon rank, proportion of reactive macerals, and extent of oxidation (5-15).
An example from the literature (5-16) is shown in Figure 5-7
(compare with Figure 5-2). The molecular weight per cross-linked
unit (M_C) has been estimated (5-16) from swelling of coals in
pyridine. Larsen has suggested that in the work of Sanada et al.
(5-16) sufficient time may not have been allowed for maximum
equilibrium swelling to have been achieved. Nevertheless, the
data are probably qualitatively correct (5-17) and in Figure 5-8
we show a comparison of M_C and coal conversion in our 3 minute
runs as a function of rank. It appears that the most reactive
coals are indeed those with the highest molecular weight per
crosslink (lowest crosslinking density).

It has also been reported (5-18) that American coals show
both a minimum in density (as determined by He sorption) and a
maximum in surface area (as determined by N_2 sorption) at 76-84%
MAF carbon. Thus in this range most of the porosity is attributed
to micropores (4-12 μ diameter) and "transitional" pores (12-300 μ
diameters). At lower ranks (76% MAF C) most of the pore volume
is in macropores (300 μ diameter). At higher ranks (85% MAF C)
microporosity predominates. Thus there may be physical reasons
for maximal initial reactivity in the middle range.

The approximate location of this maximum in conversion
potential along with lower convertibility of higher rank coals
is generally agreed upon, but there is disagreement over the
reactivity of lower rank coals (5-12).

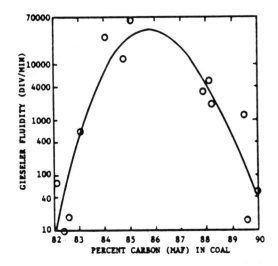

FIGURE 5-7. Maximum Fluidity vs. Rank of Coal

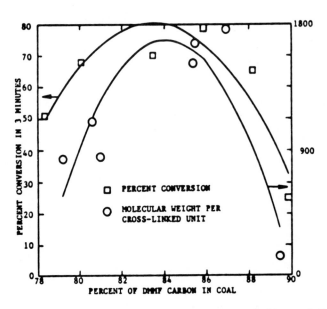

FIGURE 5-8. Relationship Between Reactivity and Rank

III.   CORRELATIONS BETWEEN COAL LIQUEFACTION
       BEHAVIOR AND COAL RANK

To shed more light on the extent of reaction as a function of
rank, the 3 minute conversion results will be compared to those
obtained at 90 minutes. The following discussions also include
coals other than the high vitrinite series, in order to extend
the rank range considered. All coals were converted with a synthetic solvent containing 8.5% tetralin, at 800°F, with H_2 for 3
or 90 minutes. In the series of runs to be discussed here, there
are two new feedstocks. The first is a horticultural peat moss
from Carlton County, Minnesota (obtained from Red Wing Peat Corporation). The second is a subbituminous coal from the Susitna Basin
in Alaska. This coal was chosen to include another subbituminous
coal from a different geological and geographical area than Wyodak-
Anderson because the structure of Wyodak-Anderson coal was so
different from other coals previously considered.
     Conversions to pyridine-solubles are plotted against coal
rank as indicated by MAF carbon in the coal in Figure 5-9. For
long times (90 minutes), conversion is high for all ranks less
than 85% MAF carbon; at higher ranks, conversion falls off sharply.
Generally lower rank coal conversions are considerably lower at
short times than long times, but in the 75-85% MAF C range they
are the same; conversion is virtually complete in 3 minutes. Very
high rank coal conversions fall off sharply from 90 to 3 minutes.
Using vitrinite reflectance as the rank parameter gives exactly
the same picture.
     The situation is slightly different for the high vitrinite
series alone. For these coals, maximum conversion appears to
occur at slightly higher rank (∿85% MAF C) and conversion falls
off with decreasing rank even at long times.
     The various product distributions and compositions for the
3 minute runs will first be discussed generally, and then followed
from 3 to 90 minutes. The products will then be reviewed as
functions of rank. Three rank regions will be examined: high
rank (>85% MAF C), low rank (<78% MAF C) and the intermediate
rank region (∿75-85% MAF C).
     At the short times, hydrogen consumption does not parallel
conversion. Apparently many reactions occur with low rank coals
which consume hydrogen but do not lead to the formation of soluble products.
     The relationship of hydrogen consumption to conversion at
short times is shown in Figure 5-10. At approximately equal conversions, low rank coals consume considerably more hydrogen than
high rank coals. This may be because for the low rank coals the
reactions which initially take place do not lead to pyridine
solubility. At long times hydrogen consumption is greatest at

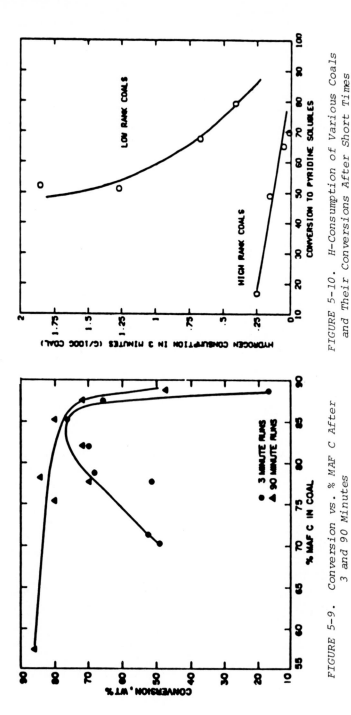

FIGURE 5-10. H-Consumption of Various Coals and Their Conversions After Short Times

FIGURE 5-9. Conversion vs. % MAF C After 3 and 90 Minutes

intermediate ranks (where there is the largest increase in hydrogen consumption with increasing time). It is also observed that
the total hydrogen content of the products (from H_2O and CH_4 to
residue) is also at a maximum for these intermediate ranks at
either 3 or 90 minutes. The maximum in hydrogen content of the
products is more pronounced at 90 minutes.

    This is partially explained by the SRC yields vs. rank. The
initial SRC yield is highest at intermediate ranks and then decreases substantially with time, producing gases, light liquids
and solvent-range distillates. This is expected to consume substantial amounts of hydrogen. The lower and higher rank coals
yield more SRC as time progresses; this additional dissolution
requires less hydrogen. The high rank coals, even at long time,
give very high SRC yields relative to conversion, i.e., very
little of other products. The peat on the other hand gave very
low SRC yield at high conversion.

    The yields of C_1, C_2-C_5, C_6-470°F, and 470-650°F products all
behave similarly, peaking at intermediate rank. The yield of
methane is about equal to C_2-C_5. The C_6-470°F and 470-650°F products show the same trends and there tends to be a little more
of the heavier liquids than the light. Using C_2-C_5 gases as an
example, Figure 5-11 shows that 650°F⁻products are minor until
about 70% coal conversion and then increase sharply at higher
conversions. There is a good correlation between hydrogen consumption and light product yields. From Figure 5-12*it can be
seen that low rank coals initially produce very little light
hydrocarbon product but ultimately produce a great deal. The
highest rank coals yield very little light hydrocarbons, even at
long times (5-27).

    Figures 5-13 and 5-14 show the CO_2 and CO yields; both are
very low at high rank and increase dramatically with decreasing
rank (as do the coal liquids' oxygen contents as shown later in
Figure 5-22). The high rank (85% MAF C) coals produce 0.5% CO_2
even at 90 minutes. The low rank coals produce 1 to 2% CO_2 in
3 minutes and 4-5% at 90 minutes; the highest rank coal produced
no detectable CO_2. At lower ranks, CO yields were higher than
at higher ranks, but never exceeded 1.5% of the coal. The
Alaskan coal gave extremely high CO and CO_2 yields at 3 minutes.

    Figure 5-15 shows the behavior of the H/C mole ratio of the
SRC vs. time. The SRCs from all coals decrease in H/C with time
except for PSOC-372 (85.1% MAF C) which does not change; this is
the coal that showed the largest decrease in SRC yield with time.
Lower rank coals show higher ratios and decrease more with time.
The H/C mole ratios of the residues also decrease with time but
there is no apparent correlation of ratio, or its rate of decrease
with rank (Figure 5-16).

Percent MAF C in coal is indicated in this and similar figures.

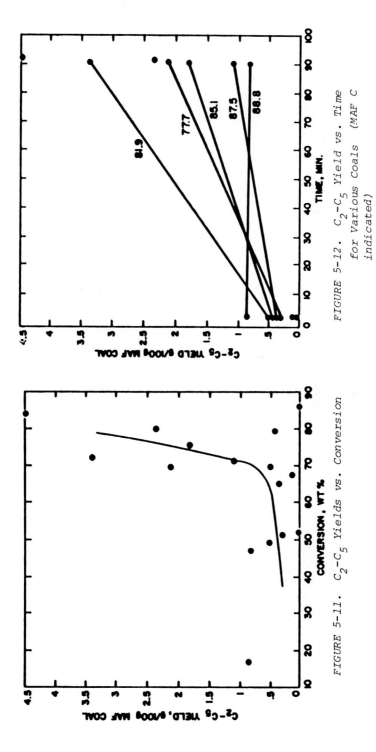

FIGURE 5-12. C_2-C_5 Yield vs. Time
for Various Coals (MAF C
indicated)

FIGURE 5-11. C_2-C_5 Yields vs. Conversion

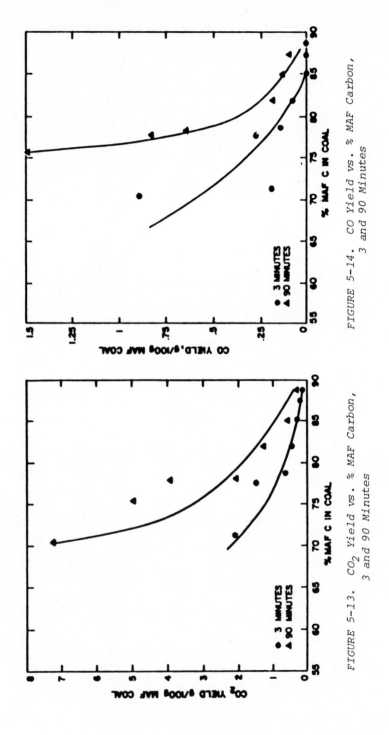

FIGURE 5-14. *CO Yield vs. % MAF Carbon, 3 and 90 Minutes*

FIGURE 5-13. *CO_2 Yield vs. % MAF Carbon, 3 and 90 Minutes*

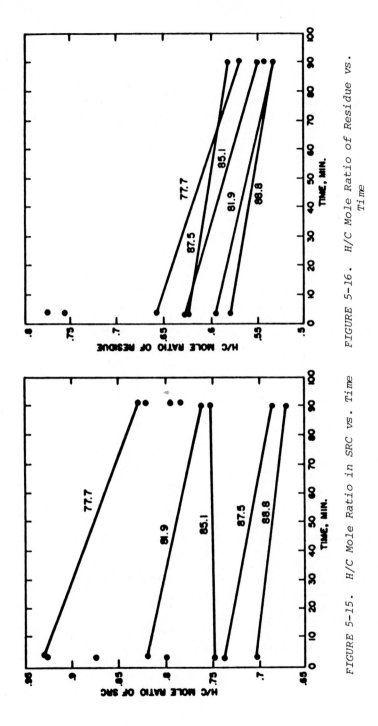

FIGURE 5-16. H/C Mole Ratio of Residue vs.
Time

FIGURE 5-15. H/C Mole Ratio in SRC vs. Time

An extensive effort was made to develop correlations of yields of SESC fractions with rank, conversion, and time. These data are generally characterized by some scatter, indicating that the chemical functionality distribution in the SRC depends on the exact nature of the coal and not just rank.

The cumulative yields of the fractions (based on MAF coal) vs. rank are shown in Figure 5-17(a) (3 minutes) and Figure 5-17 (b) (90 minutes). At the shorter time it can be seen that there is relatively little variation in the non-functional fractions (1 and 2). There is a major increase in fraction 3 (polar aromatics) at intermediate rank; it is noteworthy that these coals produce so much non-polar material in 3 minutes. Fractions 4 and 5 (monophenols and basic nitrogen heterocycles) show very little variation with rank. The high SRC yields at intermediate ranks are due substantially to high yields of fraction 3 and of the polar fractions 6-9 (asphaltols).

At 90 minutes, the lower rank coals have formed significant amounts of non-polar hydrocarbons (fractions 1 and 2) and less of fractions 6-8; fractions 3 and 9 are still the two largest fractions. The high rank coals never form appreciable amounts of fractions 1 and 2, even at longer times but do form a lot of fraction 3. They also develop more of fractions 6-8 at longer times. Fraction 9 is appreciable throughout and does not vary much.

The key fractions appear to be 6 through 9. They account for most of the high yield of SRC from intermediate ranks at short time, for the increase in yield of high ranks at longer time, and for the decrease in SRC at intermediate ranks at long time. The relative amounts of these fractions do not change as significantly as the absolute amounts.

There is some correlation at short times between fractions 6-9 and conversion or SRC yield. However, at 90 minutes the correlation is very poor. All attempts to cross-correlate the yield of any one fraction with any other, across the entire rank range, failed.

The aromaticities of most of these coals have been determined by CP-^{13}C-NMR and Fourier Transform Infrared Spectroscopy (FTIR) (5-29); these data and the conversion achieved in 3 minutes are given in Table 5-5. Conversions decrease dramatically with increasing aromaticity above 80% aromatic carbon.

The most satisfactory relationships between coal composition and reactivity at short times were related to the content of certain aliphatic components in the coal. The absolute aliphatic hydrogen content as determined by Solomon using FTIR (5-29,5-30) shows a very good linear relationship with conversion of coal in 3 minutes to pyridine-soluble materials (Figure 5-18(a)).

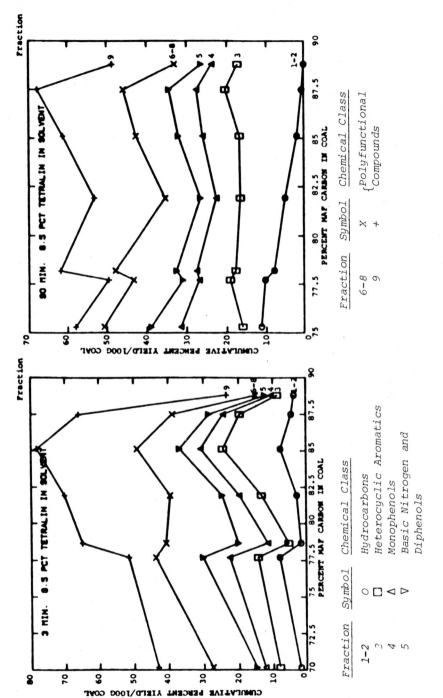

FIGURE 5-17. Cumulative Yields of SESC Fractions vs. Rank (3 and 90 Minutes)

TABLE 5-5. *Carbon and Hydrogen Type Distributions of Coals Converted at Short Times*

PSOC #	Coal	% C_{MAF}	CP-C^{13}-NMR Analyses			FTIR Analyses				% Conversion in 3 min.[c]
			% Ar[a] C	% Alkoxy C	% Al[b] C	% Ar C	% Al C	% Ar H	% Al H	
	Alaska Subbituminous	70.2	64	5	31					49
	North Dakota Lignite	71.3	69	3	28					52
	Wyoming Subbituminous	72.6	70	3	27					~60
312	Arizona Bituminous	77.7	73	6	21					51
	Illinois Bituminous	78.2	67							68
330	Pennsylvania Bituminous	81.9	73	6	21	76	24	37	58	70
372	Kentucky Bituminous	85.1	69	3	28	75	25	34	61	79
256	Pennsylvania Bituminous	87.6				79	21	44	54	65
405	Oklahoma Bituminous	88.8	84	1	15	86	14	55	44	25

A similar result was also observed using an oxidative technique developed by Deno to identify aliphatic coal constituents.
Figure 5-18(b) shows a crude correlation between hydrogens converted to carbons adjacent to aromatic rings (not including methyls
or methylene linkages between aromatics) and the yield of SRC
achieved in 3 minutes of reaction.

These results indicate that the aliphatic portion of the coal
is very important in the initial phases of coal conversion. Weak
linkages must be associated with the aliphatics in coal though
they have not as yet been completely identified. Both of the above
methods show an increase in the aromatic methyl contents of SRCs
at short times which indicates that cleavage at a benzylic carbon
is important in dissolving the coal.

There is additional evidence for the importance of aliphatic
hydrogen and its relationship to coal rank and reactivity in liquefaction. Reggel, Wender, and Raymond (5-28) studied the dehydrogenation of vitrains from a variety of coals with 1% $Pd/CA(CO_3)_2$
in refluxing phenanthridine; their results are shown in Figures
5-19(a) and (b). Coals in the rank range which rapidly give high
SRC yields are rich in hydrogen, which their technique can remove.
Furthermore, there is a distinct difference between bituminous
coals, subbituminous coals and lignites. The lower rank materials
yield less H_2 in their test; we find these to be very reactive
but slow to yield pyridine-soluble products. These workers concluded from their work "that lignites and subbituminous coals
contain some cyclic carbon structures which are neither aromatic
nor hydroaromatic; that low rank bituminous coals contain large
amounts of hydroaromatic structures; and that higher rank bituminous coals contain increasing amounts of aromatic structures."

We have also examined the fate of the oxygen in the coal as a
function of rank. There is a slight tendency for the oxygen recovery to be low, reflecting the fact that oxygen in the 257-650°F
fractions was not determined, but their oxygen content was clearly
small.

In the 3 minute conversions, the majority of the oxygen is
found in the SRC and residue in proportions approximately equal
to their yields. CO contributes very little, H_2O is 5-10% except
at very low rank, where it is a bit higher, and CO_2 gradually
increases with decreasing rank. H_2O and CO_2 generally account
for roughly comparable amounts of oxygen.

At 90 minutes the SRC and residue still account for most of
the oxygen except at the lowest rank. Less of the oxygen is
accounted for in the residue (relative to 3 minute runs) at all
ranks. The SRC accounts for less of the oxygen (relative to 3
minutes) up to about 85% MAF C; at higher ranks the SRC accounts
for more of the oxygen at longer times (reflecting increased conversion and perhaps even incorporation of p-cresol from the solvent). At 90 minutes the SRC accounts for more oxygen than the

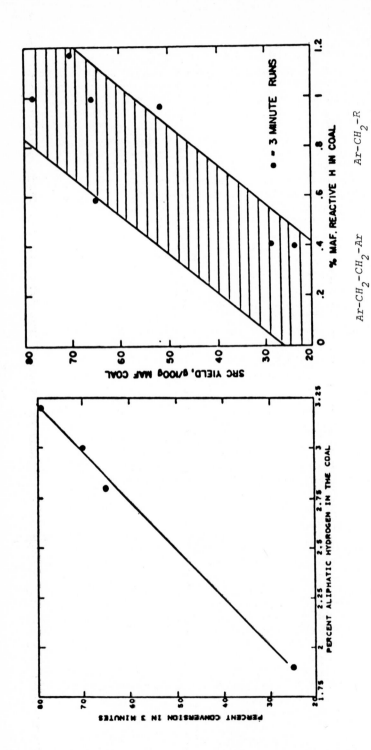

FIGURE 5-18(a). *Response of Coal Conversion to Aliphatic Hydrogen Content*

Figure 5-18(b). *Relationship Between SRC Yield and Reactive Aliphatic Hydrogen*

150

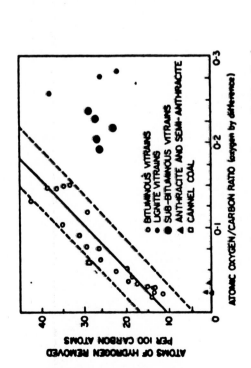

FIGURE 5-19 (a). Hydrogen Removed vs. % MAF Carbon

FIGURE 5-19 (b). Hydrogen Removed vs. Oxygen/Carbon Ratio

FIGURE 5-19. Dehydrogenation of Vitrains

residue at all ranks. Oxygen accounted for by CO rises gradually
with decreasing rank but never exceeds about 5%. Oxygen accounted
for by H_2O is very low at ranks >80% MAF C (less than it was at 3
minutes) but increases dramatically at lower ranks. Oxygen accounted for as CO_2 increases smoothly with decreasing rank. At low ranks
contributions from CO_2 and H_2O are about equal; at higher ranks CO_2
predominates.

     The contribution of various oxygen functional groups in coal
and coal products to the progress of coal liquefaction is critical-
ly important, as are the various chemical forms in which oxygen is
found in the products. We have described elsewhere the oxygen
functionalities in coals, and model compound investigations of the
possible contributions of oxygen functionality to the bond breaking
in coal dissolution. Ethers have been identified as important bonds
that can be broken, and this has been supported (5-26) by examination of the structure and products of one of the coals (PSOC-330) in
the high vitrinite series discussed in this chapter.

     An intriguing question, especially for low rank coals is, what
reactions are responsible for the high hydrogen consumption and CO_2
evolution, and what are the reactions of phenols? Figure 5-20 shows
the relationship of hydrogen consumption to CO_2 formation; we have
already shown that CO_2 evolution is greatest for low rank coals.
Figure 5-21 shows oxygen conversions at short times, and this also
increases dramatically with decreasing rank. Figure 5-21 shows the
oxygen contents of short contact time SRCs and Figure 5-22 shows
their phenol contents. The CO_2 yields were shown in Figure 5-15.
It is clear that for low ranks at short times, CO_2 yields are significant, the oxygen contents of SRCs are lower than those of the
coals, phenol as a percent of oxygen in SRCs is lower than phenol
as a percent of oxygen in the coals (and less than half of the oxygen in coals is phenolic), and total oxygen conversion is extensive.
At longer times CO_2 yields increase dramatically and the oxygen remaining in SRCs becomes more phenolic.

     Model compound studies indicate that esters (R-COOR'), and
oxylates (ROOCCOOR') are reactive functionalities that can cleave
to produce CO, CO_2 and phenols and consume hydrogen. The presence
and reactions of such compounds are consistent with all of the above
observations and perhaps this point should be investigated in the
future. The lower phenol content of the initially dissolved species
could be due either to early reversible condensation reactions of
phenols or simply to the fact that large highly-phenolic materials
would not be soluble. The formation of phenols in the initial reactions could also help explain why reactive low rank coals produce
soluble species only slowly.

     As was mentioned earlier, rank is not the sole controlling
factor of reactivity. The effects of mineral matter are discussed elsewhere and were examined in our earlier work (5-1,5-2
and 5-27). Geologic origin also appears to play a role, and the
significance of mineral matter may even be dependent upon this

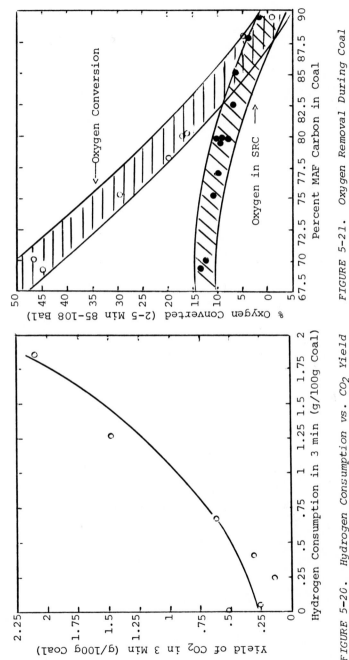

FIGURE 5-21. Oxygen Removal During Coal
Conversions

FIGURE 5-20. Hydrogen Consumption vs. CO₂ Yield
in Three Minutes

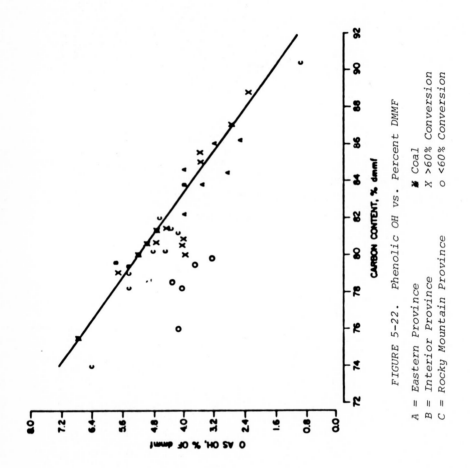

FIGURE 5-22. Phenolic OH vs. Percent DMMF

A = Eastern Province   ▲ Coal
B = Interior Province   X >60% Conversion
C = Rocky Mountain Province   o <60% Conversion

factor. Given and co-workers (5-20,5-21,5-22) have done some
important work in this area. Their reports help in particular to
explain the apparently low reactivity of PSOC-312.

These workers found that in attempting to correlate coal conversion (to ethyl-acetate-soluble material in one hour reactions
at 752°F in tetralin with no H_2) using fourteen different coal
properties, coals first had to be clustered into three different
groups for which three different regression analyses then gave
satisfactory fits. Sulfur content was the major controlling
factor that separated the three groups; factors related to rank
and petrography also contributed. Note that the coal properties
that determine the group into which a coal fell did not include
location or province and may or may not appear in the regression
equation that best predicted the conversion of that coal. The
three groups are shown in Figure 5-23.

Application of Given's criteria shows that four of the high
vitrinite series coals fall into his Group I, almost entirely
Eastern Province coals. The fifth high vitrinite coal (PSOC-312),
Wyodak-Anderson, and the Alaskan coal fall into his Group III,
almost entirely Rocky Mountain Province coals. Illinois #6
(Monterey) and Illinois #6 (Burning Star), which have rank and

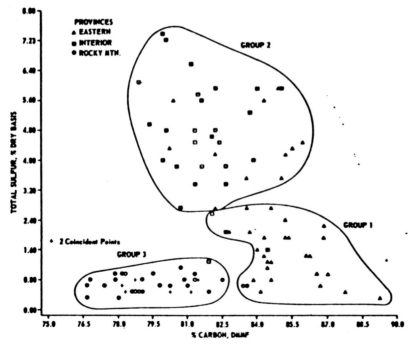

FIGURE 5-23. Clustering of Coals Using Sulfur and Carbon
Contents Only.

petrography similar to PSOC-312, fall into Group II, mostly
Eastern and Interior Province coals, which generally show higher
conversions than coals from the other groups.

   Coal PSOC-312, apparently for reasons partially related to
geologic origin (which affects but is not included in the sulfur
term that contributes heavily to the clustering into groups), be-
longs to a group of coals that behave differently in liquefaction
from the other coals in the high vitrinite series. This is why
conversion appears to decrease at lower rank for the high vitrinite series, alone, at 90 minutes but not when all of the coals
are considered together.

   For the Group III coals, conversions at 3 minutes are well
under the predicted values and conversions at 90 minutes are over
the predictions of Given. For Group II (Illinois #6 (Monterey)
and Illinois #6 (Burning Star)) the predictions are good at 3
minutes but low at 90 minutes. For the four Group I high vitrinite coals the predictions are lower than conversions at either
long or short time except for agreement between predicted and
found for the short run with the highest rank coal.

   The predictions are for ethyl acetate solubility under conditions intermediate between our long and short runs, but probably closer to the long runs because of the long time and high-
donor solvent. These predictions are probably low for the high
rank coals (except for the highest) because as was shown above
these produce large amounts of asphaltols that would be soluble
in pyridine but not in ethyl acetate. They are high for the lowest rank coals at short time because of the low rates of dissolution of these coals, but slightly low for the lower rank coals
at longer times because of the solubility differences just noted.

   Given's predictions are consistent with our findings when
rates and solubilities are taken into account and help to explain
why there can be significant differences between coals that have
similar petrography and rank.

IV.   REVIEW OF RANK EFFECTS

   The highest rank coals are relatively unreactive. The very
high rank coals (>88% MAF C) have both low ultimate reactivities
and low rates, even if they are rich in vitrinite. They produce
negligible CO, CO_2, and H_2O and very little light hydrocarbon.
Hydrogen consumption is low, especially at short times and the
H/C mole ratios of SRC and residue are also very low. These coals
have very little reactive hydrogen, are highly aromatic, and
probably contain few reactive functional groups. It has been
reported (5-23) that x-ray scattering data indicate that the
number of rings in lamella of condensed aromatic rings in coals

begins to rise rapidly with increasing rank in coals of rank>37%
MAF carbon. The SRCs they produce are composed chiefly of fractions SESC-3 and 9.

The lowest rank coals,<80% MAF C, are also slow to react, but
ultimately will reach high conversions; they show dramatic changes,
especially very large increases in light hydrocarbon and other
gases with time. SRC yields do not change much with time, but
although they have the highest H/C mole ratios in SRCs and residues, the H/C of SRCs show large decreases with time. They also
show the greatest hydrogen consumptions at short time and significant increases in hydrogen consumption with extended reaction
time. They are low in aromatic carbon. SRCs show relatively
less polyfunctionality at short times than do higher rank coals
but at longer times this effect is gone. Surprisingly, they produce the largest yields of the least functional fractions at long
times.

These coals probably have enough reactive bonds to be easily
liquefied. There is evidence in the literature (5-24) that lignite is more reactive than Wyodak-Anderson; conversion of the for-
mer is higher in a high Solvent Quality Index solvent in the EDS
process. The initial products, however, are so functional and/or
so large that they are insoluble. It has been shown (5-25) that
the hydroxyl contents of coals increase with decreasing rank; it
is, therefore, probable that this is true for initial products as
well. It has also been shown (5-26) that benzene solubility of
SRCs decreases with increasing hydroxyl content.

The coals are very reactive, as evidenced by the high hydrogen
consumptions and yields of CO, CO_2 and H_2O, even at short times.
The initial products, both soluble and insoluble in pyridine, are
also very reactive, as shown by the continuing hydrogen consumption, generation of oxygen-containing products, and formation of
low functionality SRC fractions. This might not be true in a
solvent of lower hydrogen-donor ability. In Section 7 we will show
that with less than 8.5% tetralin in a solvent, material insoluble in pyridine at short times is not very reactive. Hydrogen
donors may be rate-limiting in the runs used in the present discussion of these very reactive coals. At higher donor levels or
with better donors, formation of pyridine-soluble material might
be as fast as with the higher rank coals. This could account for
the earlier-discussed disagreement in the literature over the
ultimate conversion of low-rank coals.

The intermediate rank coals are most interesting. They reach
their ultimate conversions very quickly and give initial high
yields of SRCs that are very rich in asphaltols. At short times,
yields of products other than SRC are minor. As time progresses,
these coals produce some H_2O, CO, and CO_2, and more hydrocarbon

gases and light liquids than coals of higher and lower rank. SRC
yields decrease with time and hydrogen consumptions ultimately are
the highest. Virtually all of the products formed after the first
few minutes are produced from already pyridine-soluble material,
which is not true of coals of higher or lower rank. The intrinsic
reactivities of these coals of intermediate rank probably are not
quite as high as those of lower rank coals, so the rates are not
as prone to be limited by donor availability. However, the
initial products are soluble in pyridine because they are less
functional and probably lower in molecular weight. Therefore,
the coals "appear" to be more reactive. Further reaction gives
more light hydrocarbon gases and liquids. There must be more O,
N and S in rings than in the lower rank coals, because less of
SESC fractions 1 and 2 are ultimately formed.

Thus the overall picture is that intrinsic reactivity
actually decreases with increasing rank over the entire rank
range. However, the common criterion for reactivity is the production of soluble material and low rank coals require good
sources of hydrogen, so some workers have concluded that these
coals are less reactive than those of intermediate rank.

REFERENCES

5- 1.   D. D. Whitehurst, M. Farcasiu, and T. O. Mitchell, "The
          Nature and Origin of Asphaltenes in Processed Coals",
          EPRI Report AF-252, First Annual Report Under Project
          RP-410, February 1976.

5- 2.   D. D. Whitehurst, M. Farcasiu, T. O. Mitchell and J. J.
          Dickert, Jr., "The Nature and Origin of Asphaltenes in
          Processed Coals", EPRI Report AF-480, Second Annual
          Report Under Project RP-410-1, July 1977.

5- 3.   C. H. Fisher, G. C. Sprunk, A. Eisner, H. J. O'Donnel,
          L. Clarke and H. H. Storch, U. S. Bureau of Mines
          Technical Paper 642 (1942).

5- 4.   H. H. Storch, C. H. Fisher, C. O. Hawk, and A. Eisner,
          U. S. Bureau of Mines Technical Paper 654 (1943).

5- 5.   P. H. Given, D. C. Cronauer, W. Spackman, H. L. Lovel,
          A. Davis and B. Biswas, FUEL, 54, 34 (1975).

5- 6.   H. H. Oelert and R. Sickmann, FUEL, 55, 39 (1976).

5- 7.   D. W. VanKrevelen, "Coal, Typology-Chemistry-Physics-
          Constitution", Elsevier Publishing Co., New York,
          1961, p. 119.

5- 8.   W. Francis, "Coal, It's Formation and Composition",
          Edward Arnold (Publisher) Ltd., London, 1961, pp. 359,
          419.

5- 9.   D. D. Whitehurst, unpublished work.

5-10(a). R. C. Neavel, FUEL, 55, 237 (1976).

    (b). R. C. Neavel, Proceedings of the Coal Agglomeration and
          Conversion Symposium, West Virginia Geological and
          Economic Survey and Coal Research Bureau of West
          Virginia University, Morgantown, West Virginia, May 1975.

5-11.   G. P. Curran, R. T. Struck and E. Gorin, I&EC Process
          Design and Development, 6, 166 (1967).

5-12.   A. Davis, W. Spackman, and P. H. Given, Energy Sources,
          3 (1) 55 (1976).

5-13(a). Characterization of Mineral Matter in Coals and Coal
          Liquefaction Residues", Annual Report to EPRI (Research
          Project 366-1) for 1978, The Pennsylvania State
          University, 1979.

    (b). Characterization of Mineral Matter in Coals and Coal
          Liquefaction Residues", Annual Report to EPRI (Research
          Project 366-1) for 1977, The Pennsylvania State
          University, June 1978.

5-14.   P. H. Given et al., "The Relation of Coal Character-
          istics to Liquefaction Behavior", Quarterly Technical
          Progress Report October-December 1976 (October 1977),
          FE-2494-2 to U. S. D. O. E. under contract EX-76-C-01-
          2494.

5-15.   W. Spackman et al., "The Characteristics of American
        Coals in Relation to Their Conversion into Clean Energy
        Fuels", Quarterly Technical Progress Report October-
        December 1977 (March 1978), FE-2030-10, to U. S. D. O. E.
        under contract EX-76-C-01-2030.

5-16.   Y. Sanada and H. Honda, FUEL, 45, 295 (1966).

5-17.   J. W. Larsen, D. D. Whitehurst, T. O. Mitchell. To be
        published.

5-18.   H. Gan, S. P. Nandi, and P. L. Walker, Jr., FUEL, 51,
        272 (1972).

5-19.   N. C. Deno et al., "Coal Structure and Coal Liquefaction",
        Final Report to EPRI, Project RP-779-16, 1979.

5-20.   M. Abdel-Baset, R. F. Yarzab and P. H. Given, FUEL, 57,
        89 (1978).

5-21.   M. Abdel-Baset, R. F. Yarzab and P. H. Given, FUEL, 57,
        95 (1978).

5-22.   R. F. Yarzab, P. H. Given and A. Rabinovich, "Relation
        of Coal Characteristics to Coal Liquefaction Behavior:
        Cluster Analyses for Characteristics of 104 Coals",
        Technical Report Number 3 to U. S. D. O. E. under
        contract EX-76-C-01-2494, August 1978.

5-23.   A. W. Scaroni and R. H. Essenhigh, ACS Division of Fuel
        Chemistry Preprints, 23 (4), 124, (1978).

5-24.   EDS Coal Liquefaction Progress Development Phases IIIB/
        IV, Annual Technical Progress Report for the period
        July 1, 1977 to June 30, 1978, Prepared for U. S. D. O. E.
        under Cooperative Agreement EF-77-A-01-2893 (September
        1978).

5-25.   P. H. Given et al., "The Relation of Coal Character-
        istics to Coal Liquefaction Behavior", Report Number 4
        from The Pennsylvania State University to NSF, December
        1, 1975.

5-26.   A. J. Szladlow and P. H. Given, ACS Division of Fuel
        Chemistry Preprints, 23 (4), 161 (1978).

5-27.   D. D. Whitehurst, T. O. Mitchell, M. Farcasiu and J. J.
        Dickert, Jr., "The Nature and Origin of Asphaltenes in
        Processed Coals", Third Annual Report and Overall
        Summary Under EPRI Research Project 410-1, March 1977-
        January 1979.

5-28.   L. Reggel, I. Wender and R. Raymond, "Catalytic De-
        hydrogenation of Coal III. Hydrogen Evolution as a
        Function of Rank", Fuel (Lond.) 47, 373 (1968).

5-29.   D.D. Whitehurst and P.R. Solomon, to be published in Fuel.

5-30.   P.R. Solomon, "Relation Between Coal Structure and
        Thermal Decomposition Products", Preprints Fuel Division,
        ACS/CJS Chemical Congress, Honolulu, Hawaii, March 1979,
        p. 184.

Chapter 6

CATALYTIC EFFECTS OF MINERAL MATTER ON COAL LIQUEFACTION

Many scientists have reported the benefits of intrinsic mineral matter for catalyzing coal conversion reactions (6-1). The
observed catalytic activity is usually ascribed to pyrite, or,
pyrrhotite (6-1). Relatively little work, however, has been done
on the mechanisms of the possible synergism of acid and metal activity, or on the relationship of such synergism to solvent rehydrogenation. Some aspects of these mechanisms will be discussed
in this section.

In a recent study (6-2) on the hydrogenation of a high-vitrinite Indian coal (North Assam) in the absence of a solvent, the catalytic effect of mineral matter was studied by characterizing the
coal ash and by adding specific minerals. The best correlation to
activity was found using (organic plus pyritic) sulfur. Other
materials -- iron, titanium and kaolinite (the prevalent clay) --
also correlated with coal conversion to benzene-soluble products.
Iron pyrite was suspected as the active form of iron but conversion also increased with the addition of sulfur or titanium hy-
droxide.

In another recent report (6-3), pyrite addition increased the
pyridine solubility of four German coals (after 2 hours in methylnaphthalene at 752°F under 3000 psi of hydrogen). Samples of a
coal enriched in mineral matter were more extensively converted.

For both low and high pyrite-content coals, mineral matter
affects the interactions of solvent components; in addition, the
rates of solvent-solvent hydrogen-transfer reactions are higher in
the presence of coal than in blank reactions. In our studies involving iron pyrite and synthetic solvent mixtures, all solvent
reactions are accelerated by pyrite, but the relative rates do not
change. The reactions studied included tetralin isomerization to
methylindan, its dehydrogenation to naphthalene, and its hydrogen

transfer to methylnaphthalene. Since pyrite reduces nearly quantitatively to H_2S and FeS during the first two minutes, the increases in rates of solvent-solvent interactions are probably not due to iron sulfides alone; hydrogen sulfide may act as a radical chain-transfer agent.

Other possible catalytic materials are acid clays. A natural bentonite clay had relatively little effect on solvent-solvent interaction, but it also had a very low acid activity.

Several studies have been made of the catalytic activities of coal residues from autoclave reactions. For acid-catalyzed t-butyl acetate decompositions (6-4, 6-5), the highest activities are observed with the parent coals rather than with the autoclave residues. Activity (based on the total residue or ash content) decreased with increasing (autoclave) contact time; the activities of different coals varied considerably.

Studies of dehydrogenation, hydrogenation and dehydroxylation in the system acetone $\rightleftarrows$ isopropyl alcohol $\rightleftarrows$ propylene (6-6) show that for isopropyl alcohol conversion, iron pyrite may not be the major catalytic agent in autoclave residues. The activities on a per-gram-ash basis are highest for residues from short autoclave runs. In addition, Wyodak-Anderson residue was more active on a per-gram-ash or per-gram-residue basis than residues from bituminous coal. Little correlation was noted between surface area and activity for the bituminous coals.

Attempts at MRDC to develop an understanding of the influence of coal mineral matter on catalysis in the SRC process concentrated on three areas: the relative rates of coal hydrogenation and solvent rehydrogenation, the importance of thermodynamic constraints, and the effect of certain classes of compounds on a solvent's ability to donate hydrogen.

Three major techniques were involved. First, blank autoclave reactions were run using solvents, hydrogen gas and minerals representative of those in coal to determine their effects on reactions such as hydrogenation, dehydrogenation, cracking, isomerization, and hydrogen transfer. Second, the mineral matter in coals was characterized in detail. Third, coals were treated to remove some or all of the mineral matter and then converted under standard conditions.

Specific catalytic studies included:

- Hydrogen transfer between two and three-
  membered ring systems.

- CoMo-Al_2O_3 catalysis versus reactor
  solid catalysis.

- Rates of dehydroxylation of mono- and
  polyaromatic phenols.

I.   GENERAL COMMENTS ON CATALYTIC REACTIONS OF RECYCLE SOLVENTS

The compositional changes which occur in a typical recycle solvent during the course of the SRC process are discussed in Chapter 9 and Reference 6-4. Briefly, hydrogen is rapidly transferred from hydrogen donors in the solvent to (thermal) fragments of coal in the preheater and in the dissolver. Hydrogen gas is consumed in small amounts in the preheater and to a larger degree in the dissolver, primarily in the production of methane and/or new solvent and in rehydrogenation of the H-donor-depleted recycle solvent. These reactions of gaseous hydrogen are believed to be catalyzed by the solids (particularly iron solids) which accumulate in and occupy up to 30-40% of the volume of the dissolver in the SRC-I process (6-8).

Reactions such as these are required to achieve a steady-state recycle solvent composition; otherwise the quality of the solvent would continually decrease. Coal dissolution would then become limited and irreversible char formation would be accentuated. The dissolver solids appear to be only modest catalysts, since typical process-derived recycle solvents contain far from the thermodynamically-allowed equilibrium concentrations of hydroaromatics.

Other reactions occur concurrently with solvent rehydrogenation in the dissolver. The most important, of course, involve upgrading SRC by removing heteroatoms and producing additional solvent. Several possible classes of reactions are shown in Figure 6-1.

The rates of catalytic hydrogenation of various aromatic compounds by cobalt-molybdenum oxides appear to be in the following order. (Data were taken from 6-9, 6-10, 6-11, and 6-12 and are shown in Figure 6-2.)

Phenols $\gg$ 3-membered rings $\approx$ 2-membered rings > biphenyls $\approx$ monoaromatic rings

By contrast, the reverse of this reaction (dehydrogenation) exhibits a somewhat different order:

Dihydro 3-membered rings $\gg$ tetra or octa hydro 3-membered rings > 2-membered rings > cyclohexylbenzenes $\gg$ cyclohexanes

Thus, under typical liquefaction conditions, the two membered ring systems could have the highest concentrations of the hydroaromatics capable of donating hydrogen to coal. The fact that dissolver solids appear to possess little catalytic hydrogenation-dehydrogenation activity raises a number of questions that will be discussed below.

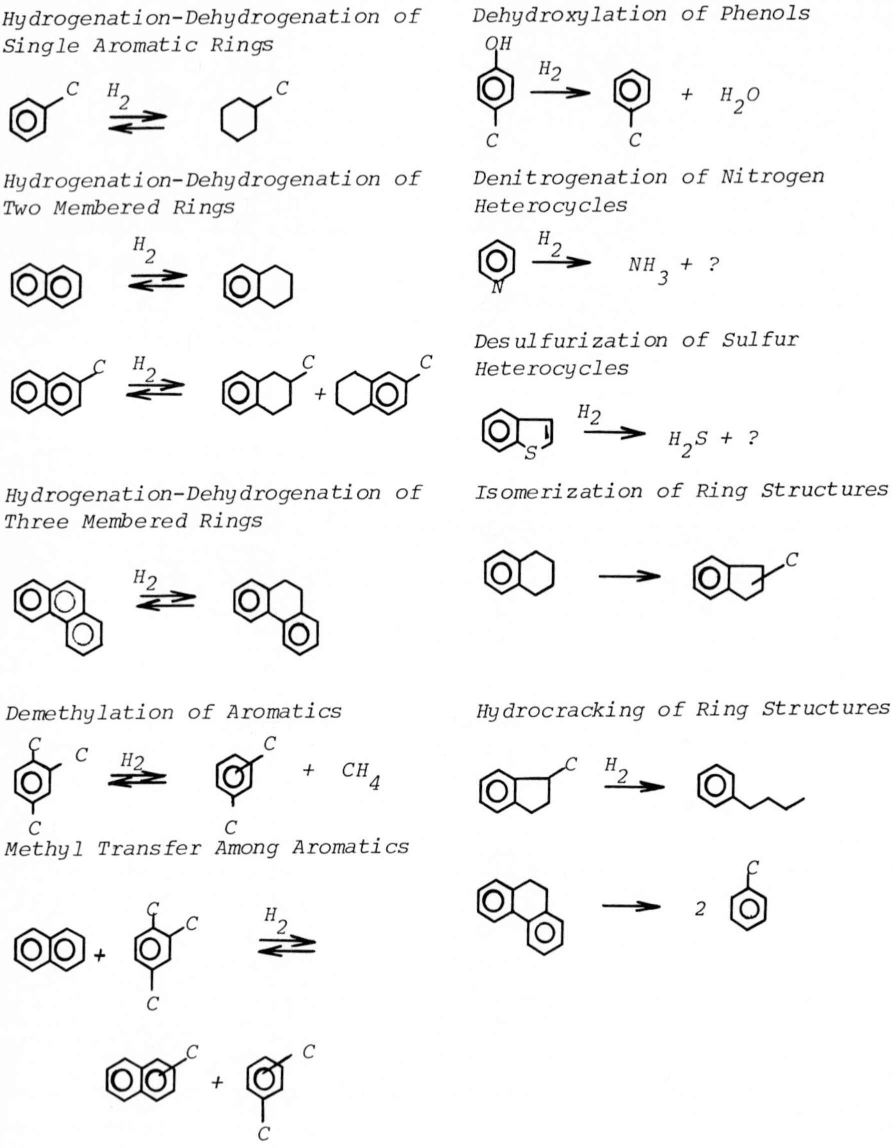

Hydrogenation-Dehydrogenation of Single Aromatic Rings

H_2

Dehydroxylation of Phenols

H_2 + H_2O

Hydrogenation-Dehydrogenation of Two Membered Rings

H_2

Denitrogenation of Nitrogen Heterocycles

H_2 NH_3 + ?

H_2 +

Desulfurization of Sulfur Heterocycles

H_2 H_2S + ?

Hydrogenation-Dehydrogenation of Three Membered Rings

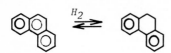

H_2

Isomerization of Ring Structures

Demethylation of Aromatics

H_2 + CH_4

Hydrocracking of Ring Structures

H_2

Methyl Transfer Among Aromatics

H_2

2

+   +

FIGURE 6-1.   Reactions which alter solvent composition.

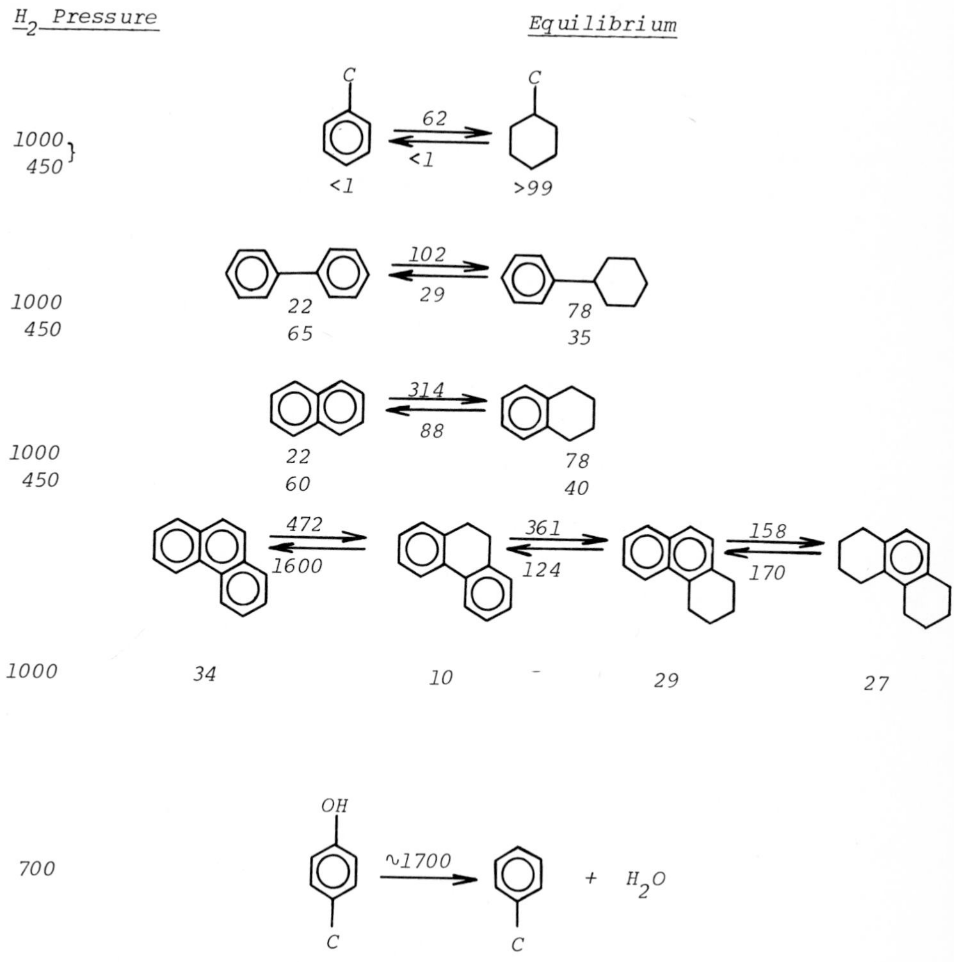

FIGURE 6-2.   Hydrogenation of Representative Solvent Compounds
              (6-9, 6-10, 6-11, 6-12).

NOTE: The numbers above the arrows show the rate constants over
Co/Mo catalysts at 800°F and ~1000 psi H_2.   The numbers below the
structures indicate the equilibrium concentrations of the indicated species at the indicated H_2 pressures.

A.   THERMODYNAMIC EQUILIBRIA OF TETRALIN-NAPHTHALENE

If the two-membered ring hydroaromatics are important in the SRC process, the thermodynamic equilibrium ratios of tetralin and naphthalene under typical process conditions are very relevant. Several reports are available with information on this ratio (6-9, 6-10).   The most thorough treatment, by Frye (6-9), however, had a limited temperature range.   We have updated his predictive equations using data reported by other authors (6-10, 6-13, 6-14). Considering all data, the best linear fit of the equilibrium constant vs. temperature, and the equations which result are as follows:

$$K_p = \frac{(\text{moles tetralin})}{(\text{moles naphthalene}) \ P_{H_2} + 3.30 \times 10^{-4} P_{H_2})^2}$$

where P_{H_2} is the H_2 partial pressure in atmospheres.

$$K_p = -13.3689 + 7158.47T$$

and T is in degrees Kelvin.

Values of the percent tetralin in the tetralin-naphthalene pair at 750, 800, 850 and 900°F from 0 to 2000 psi H, are presented graphically in Figure 6-3, based on the above equations.

At the Wilsonville PDU the partial pressure of H_2 typically varies from 500 to 1400 psi (6-7) which corresponds to equilibrium concentrations of tetralin of 30 and 75%, respectively, at temperatures of 830-860°F.   The concentration of tetralin in Wilsonville recycle solvents is far from equilibrium; typically the concentraticn of tetralin in the tetralin-naphthalene pair, is 16% to 30%. At low H_2 partial pressure, the H_2 consumption decreases; this is consistent with less solvent rehydrogenation.

This study of the tetralin-naphthalene pair is only a simplification.   Other pairs of hydroaromatic-aromatic compounds are present in a recycle solvent.   The thermodynamic equilibria of other such pairs should be calculated.   The ratio of fractions RSMC-2/ RSMC-3 (tetralin-type/naphthalene-type hydrocarbons) is likely to give a better understanding of the overall situation.

9,10-Dihydrophenanthrene is a much more reactive H-donor than tetralin is, but little evidence has been found for its presence in typical SRC solvents (6-7).   It is expected that phenanthrene will hydrogenate more readily than naphthalene (see Figure 6-2) but its high reactivity may prevent high concentrations.

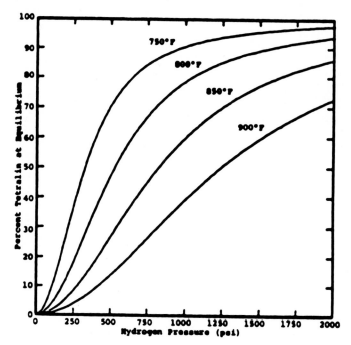

FIGURE 6-3. Tetralin/Naphthalene Equilibrium

B. CATALYSIS OF HYDROGEN TRANSFER BETWEEN TWO- AND THREE-MEMBERED RING SYSTEMS

If mineral matter can catalyze hydrogen transfer between three or larger membered ring systems and two-membered ring systems, the more easily hydrogenated larger rings may transfer hydrogen to form the tetralin homologs observed in typical SRC solvents. It is noteworthy that in our model compound experiments pyrites did not promote this type of reaction. For example, in reactions with H_2, 2-methylnaphthalene, and 9,10-dihydrophenanthrene in sealed tubes at 800°F for about one hour, with or without pyrite or bentonite, the only appreciable reaction was the dehydrogenation of 9,10-dihydrophenanthrene to phenanthrene. Acid-washed montmorillonite, on the other hand, promoted not only dehydrogenation but also isomerization and hydrogenation of the 2-methylnaphthalene. Since at the low H_2 pressures employed, the hydrogenation of naphthalenes by H_2 is not thermodynamically favored, the reaction most probably involved hydrogen transfer.

C.   CATALYSIS OF MODEL SOLVENT REACTIONS BY DISSOLVER SOLIDS --
COMPARISONS WITH CoMo-Al$_2$O$_3$ CATALYZED AND THERMAL BLANK REACTIONS

It has been claimed that reactor solids which build up in
dissolvers exhibit significant catalytic activity, especially for
recycle solvent regeneration and SRC upgrading. This was tested
by comparing the activity of a commercial catalyst with that of a
high-iron (about 15% Fe$_2$O$_3$) Wilsonville dissolver solid, formed
during Illinois #6 (Monterey) coal conversion (6-15). The
reactions were run with model compounds in a stirred autoclave at
800°F under H$_2$.

The dissolver solids showed no obvious catalytic effect on
hydrogenation-dehydrogenation, isomerization, dealkylation, or
hydrogen transfer; there was slight catalysis of dehydroxylation
of single-ring phenols. CoMo/Al$_2$O$_3$ on the other hand catalyzed
all these reactions, especially dehydrogenation and dehydroxylation; aromatic-hydroaromatic equilibrium was established.

Dissolver solids must possess some catalytic activity
(however low) since in process demonstration units hydroaromatics
are found in significant steady-state concentrations and most
workers believe they are formed catalytically. The observed
inactivity in the cited autoclave experiments may have been due
to the absence of sulfur (which would have kept the catalyst
sulfided). In studies of catalytic hydrogenation and heteroatom
removal, using model compound feeds in fixed-bed reactors,
dissolver solids were found to have measurable activities for all
reactions when H$_2$S was present (6-16).

In the dissolver-solids autoclave experiments, evidence was
also sought for the possibility that a multi-ring aromatic-ring
system like phenanthrene-hydrophenanthrene might be hydrogenated
catalytically, then donate hydrogen to tetralin-naphthalene, then
itself be readily hydrogenated by gas phase H$_2$. This was not
observed, but in the presence of phenanthrene the selectivity of
methyl transfers among trimethyl benzene species was altered.
Multi-ring aromatic-hydroaromatic systems therefore can affect and
possibly can catalyze reactions of other species. Some phenol
dehydroxylation was observed with the dissolver solids; this
reaction could be useful as a test for the activity of these
obviously very low activity materials.

D.   METHYL TRANSFER AND HYDROCRACKING

At the severe conditions of liquefaction (800°F, 800-2000 psi
H$_2$), 1,2,4-trimethylbenzene (TMB) is slowly demethylated in the
presence of a CoMo/Al$_2$O$_3$ catalyst. Xylene is the primary product,
with some demethylation to toluene. The rate of this reaction is
enhanced slightly when phenanthrene is injected into the mixture.

Although phenanthrene shows only a slight effect on TMB de-methylation, it has a very marked negative effect on methyl trans-fers among TMB species. In the absence of phenanthrene, durene is formed fairly rapidly. On addition of phenanthrene, however, durene is no longer formed. Phenanthrene in addition to being hydrogen-ated and hydrocracked appears to accept methyl groups.

One of the predominant products of the hydrocracking of phen-anthrene (over CoMo/Al$_2$O$_3$ catalyst) appears to be toluene. Other hydrocracking reactions have also been observed. When CoMo/Al$_2$O$_3$ is used, butylbenzene is a major product of tetralin degradation. This product is not observed either thermally or when only dis-solver solids are present. Butylbenzene is not a direct product of tetralin but is formed from methylindan, which is formed therm-ally and/or catalytically. Methylindan is consumed about five times faster than it is formed.

The above reactions could be useful indicators of catalytic activity and could provide some useful information on the activity of dissolver solids.

E. DEHYDROXYLATION OF MULTIAROMATIC RING PHENOLS

In coal conversions without added dissolver solids, compari-sons of coal and SRC phenol contents show that about half of the phenols are removed at relatively short times.

Single aromatic ring phenols are very stable under the con-ditions of thermal coal liquefaction, although very slow catalysis of the dehydroxylation of simple phenols (such as p-cresol) is ob-served with large concentrations of dissolver solids. But even the catalyzed rates of dehydroxylation cannot account for the observed rates of defunctionalization (phenolic-OH loss) which occur during coal liquefaction.

This discrepancy led to the study of dehydroxylation of more complex phenols such as 1- and 2-hydroxynaphthalene. These stud-ies were conducted in autoclaves at 800°F and about 1500 psi for 90 minutes, with and without hydroaromatics, hydrogen gas or coal.

Dehydroxylation of naphthols occurs only to a small extent when only H$_2$ gas or a hydrogen donor and H$_2$ gas are present. When coal is present, however, naphthols dehydroxylate quite readily. Thus coal or the minerals in coal are catalysts for this reaction. Only multi-aromatic ring phenols are dehydroxylated to produce hydrocarbons during liquefaction since in cases where hydroaro-matic phenols are also present, dehydrogenation occurs before de-hydroxylation. Intrinsic coal mineral matter is a sufficient cata-lyst to promote dehydroxylation of multi-ring phenols and massive amounts of reactor solids would accelerate dehydroxylation of both single and multi-aromatic ring phenols.

II.   CONVERSIONS OF SELECTIVELY DEMINERALIZED COALS

Another approach to the study of the effects of mineral matter on coal conversion is to alter the coal mineral content since pyrites, clays, carbonates, and exchangeable ions can be removed selectively from coals. The mildest possible conditions are necessary to minimize alteration of the organic components.

In one series of experiments designed to remove all mineral matter except pyrites, Wyodak-Anderson, Illinois #6 (Monterey), and Illinois #6 (Burning Star) coals were exhaustively treated with HCl, then HF and again with HCl. (The method used was essentially as described by Radmacher and Mohrhauer (6-17) and later by Bishop and Ward (6-18), except for longer treatment at lower temperatures to minimize modification of the organic components.)

This demineralization did not grossly alter the organic portion of the original coal, as shown by exhaustive Soxhlet pyridine-extraction of the demineralized Wyodak-Anderson coal, which yielded 15% soluble products compared to about 12% for the untreated coal. Fourier Transform Infrared Spectroscopy also showed little change in the organic constituents (by comparison of untreated and demineralized Wyodak-Anderson coal) (6-19).

The inorganic mineral matter in the treated and untreated coals was examined by ashing and subsequent emission spectrographic analysis. Iron was the only metal present in major amounts (greater than 10% of ash) in all the treated and untreated samples (even Wyodak-Anderson) but silicon (in Illinois #6 (Burning Star) and (Monterey)) and aluminum (in Illinois #6 (Burning Star)) were the elements most affected by the treatment. Among the exchangeable cations initially present in larger amounts, potassium was significantly removed from the Illinois #6 (Monterey) coal and calcium from the Wyodak-Anderson; neither was changed as much in the other coals. Sodium was generally less affected. The remaining elements, present in smaller amounts, appear to be uniformly affected by the acid treatments, as their relative concentrations remained the same. The absolute amounts of ash, of course, were greatly reduced. The acid treatment did not appreciably change the H/C ratios, nor did it appear to have removed pyritic sulfur; sulfate was removed.

In another series of experiments, coals were treated with HCl only, removing exchangeable ions and partially demineralizing the coals by removing carbonates and similar salts. Pyrites and the acid forms of clays remained in the treated coal. The partial demineralization procedure is essentially the same as the HCl portion of the HCl-HF-HCl treatment used earlier.

Finally, pyrite was selectively removed from the parent coals or the HCl-HF-HCl treated Illinois #6 (Burning Star) coal by the Meyers process of TRW to give coals which were either pyrite-deficient or coals in which all mineral matter had been removed (6-20). The effect of demineralization on the liquefaction behavior of the above coals will be discussed below in two parts; effects on short time conversions and effects on long time conversions.

A. EFFECTS AT SHORT TIME

During the initial stages of coal liquefaction (<5 minutes) a portion of the coal rapidly becomes soluble. With Western subbituminous coals this portion is considerably less than for Eastern bituminous coals. By contrast, if the minerals are removed from the subbituminous coal (Wyodak-Anderson) the rate of solubilization is markedly increased and approaches the values obtained for bituminous coals (see Table 6-1).

TABLE 6-1.   *Liquefaction Behavior of Demineralized Coals*
(800°F, ~1000 psi H_2)

Coal	Coal Treatment	Liquefaction Time	Wt % SRC Yield
Wyodak	None	1.3	38.5
Wyodak	None	3.6	21.5[a]
Wyodak	HCl	3.5	63[a]
Wyodak	HCl/HF/HCl	3.0	68.5[a]
Burning Star	None	2.0	61.4
Burning Star	HCl/HF/HCl	3.0	65.4[a]

[a] *SRC Yield obtained by workup of withdrawn samples.*

Table 6-1 shows these results as well as comparisons of the
liquefaction behavior of several other demineralized and parent
coals.

The table shows that even a rather mild treatment such as HCl
extraction markedly accelerates the rate of solubilization of
Wyodak-Anderson coal. Illinois #6 (Burning Star) coal on the other
hand showed little increase in dissolution rate. This bituminous
coal, however, was highly reactive even without treatment.

Several explanations are possible for the rate increase on
demineralization but the most preferred is that the physical mode
of coal conversion was altered by removal of certain cations
(especially Ca^{+2}).

It has been shown (6-21) that coals which dissolve rapidly are
highly swelling coals. Wyodak-Anderson coal which is non-swelling
appears to dissolve via a shell-progressive mechanism (6-15). Removal of basic cations such as Ca^{+2} could improve swelling properties. The inverse is known (6-22); that is, if a swelling coal is
impregnated with as little as 5% of a basic oxide its swelling
properties are destroyed.

The generality of increased reactivity due to demineralization is not known but if the above interpretation is correct it
would be most beneficial for non-swelling coals of low rank.

B.   EFFECTS AT LONG TIMES

In contrast to the beneficial effects noted at short times,
little benefit was found for demineralization of coals in long
duration liquefaction experiments. In fact, for all coals studied
"conversions" and SRC yields were lower for demineralized coals
than for the parent coals (see Table 6-2).

At these extended liquefaction times, the HCl-HF-HCl treated
Illinois #6 (Burning Star) coal showed a lower apparent conversion,
slightly lower SRC yield, much lower solvent yield, greater hydrogen consumption, and higher gas production than untreated coal.
These results are consistent with increased reactivity of the coal
and/or its products or increased activity of the mineral matter.
The results are similar to those with HCl-treated Wyodak-Anderson
in which it was observed that treated coal gave lower SRC yields
but higher solvent and gas yields. Hydrogen consumption was also
greater. The final Wyodak-Anderson SRCs contained slightly more
oils and slightly less oxygen but the SESC distributions were very
similar. The lower SRC yields apparently did not result from the
selective loss (or non-production) of particular SRC components.
The higher activity results in an undesirable product distribution
with higher gas production at the expense primarily of solvent and
light liquid, which is consistent with the high hydrogen consumption.

TABLE 6-2 . Effect of Mineral Matter Removal on Reactivity (800°F, 90 minutes, 1500 psig H_2)

AC No.	Coal	Treatment	% Ash	Tetralin	H_2 Cons. from Gas (g/100g coal)	Conversion(%)	SRC (%)
160	ILL#6 BS	As Received	9.58	No	2.06	69.3	58.2
155	ILL#6 BS	DeFeS$_2$	10.73	No	1.01	46	58.7
134	ILL#6 BS	As Received	9.6	Yes[b]	.60	88.5	72.3
128[a]	ILL#6 BS	HCl-HF	2.02	Yes	.73	81.1	64.5
159	Wyodak	As Received	4.06	No	1.55	52.7	48.4
158	Wyodak	As Received	4.06	Yes	.22	88.2	63.3
71	Wyodak	HCl-HF	0.44	Yes	-.25	81.0	46.8
125	Wyodak	HCl	5.0	Yes	0.34	82.5	54.4

[a] 137 Minutes.

[b] Solvent contained 6-hydroxytetralin and unsubstituted tetralin.

A clear demonstration of how intrinsic minerals (especially pyrite) can play an important role in coal liquefaction is found by comparing the behaviors of unaltered and pyrite-free Illinois #6 (Burning Star) coal in a solvent which contains no hydroaromatic H-donors. Both coals were treated in identical reactions at 800°F, about 1000 psi H_2, and 90 minutes, in a solvent mixture containing 2% picoline, 18% p-cresol, and 80% 2-methylnaphthalene (no tetralin).

Both SRC yield and coal conversion paralleled the total consumption of hydrogen (Table 6-2). When hydrogen could not be supplied from the liquid phase via a hydroaromatic, then H_2 gas provided the needed hydrogen. This reaction, however, appeared to be catalyzed by intrinsic coal minerals as the mineral-free coal consumed less hydrogen gas and converted to a lesser extent. Thus, demineralization, especially pyrite removal, is detrimental to coal liquefaction when hydroaromatics are limited in the solvent.

IV. GENERAL OBSERVATIONS OF MINERAL MATTER EFFECTS IN COAL CONVERSION

When coal was reacted in the presence of H_2 with a synthetic solvent containing tetralin and 9,10-dihydrophenanthrene, the solvent appeared to approach a steady-state value of about 7% of 9,10-dihydrophenanthrene in the phenanthrene products. In addition, the concentration of 1,2,3,4-tetrahydrophenanthrene continually increased to a value of 3.7% of the phenanthrene products. These observations indicated that some hydrogenation catalysis was operative and coal minerals may be responsible.

Wyodak-Anderson coal when treated with HCl or HCl-HF became depleted in some mineral components, as described above. These treatments caused a significant drop in final conversion (88 to 81 %) in ultimate SRC yield (70 to 47 %), and in the hydrogen consumption. It should be noted that Wyodak-Anderson coal contained relatively little pyrite but catalysis of hydrogen gas reactions did respond to total ash content. Such behavior would indicate that even ion-exchangeable iron may have catalytic properties (see Figures 6-4, 6-5, and 6-6).

In summary there are clearly effects of coal mineral matter on the progress of liquefaction. Mineral matter catalyzed hydrogen gas consumption and other reactions of coal and its products. It must also aid in solvent rehydrogenation but its activity is low. Acid demineralization, especially for subbituminous coal, increases coal reactivity but decreases conversions and SRC yields at long coal conversion times because of increases in both regressive and forward reactions.

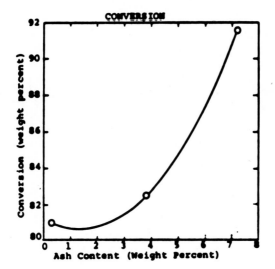

FIGURE 6-4.   Wyodak Coal Conversion vs. Ash Content

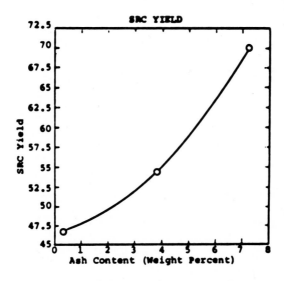

FIGURE 6-5.   Wyodak SRC Yield vs. Ash Content

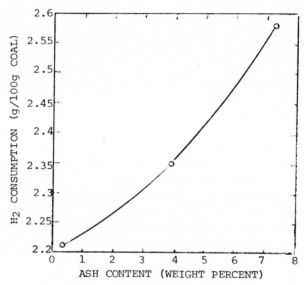

FIGURE 6-6. *Wyodak H-Consumption vs. Ash Content*

REFERENCES

6- 1.   Conference on Mineral Matter Effects in Coal Liquefaction
        Sponsored by DOE/Sandia Laboratories, Albuquerque, N.M.,
        June 14-15, 1978.
6- 2.   D. K. Mukherjee and P. B. Chowdhury, FUEL, 55, 4, (1976).
6- 3.   W. Hodek and L. Kuhn, Erdol und Kohle-Erdgas-Petrochemie,
        30(2), 85 (1977).
6- 4.   D. D. Whitehurst, M. Farcasiu, T. O. Mitchell, and J. J.
        Dickert, Jr., "The Nature and Origin of Asphaltenes in
        Processed Coals", EPRI Report AF-480, Second Annual Report
        Under Project RP-410-1, July 1977.
6- 5.   T. P. Goldstein, Twenty-third Clay Minerals Conference,
        Cleveland, Ohio, October 6, 1974.
6- 6.   D. Ollis and J. P. Joly, Chemical Engineering Department,
        Princeton University, unpublished results.
6- 7.   Wilsonville SRC Technical Report Number 8, from Southern
        Services, Inc. to EPRI, April 30, 1976.
6- 8.   W. Weber, EPRI Symposium on Reactor Solids, The Pennsylvania
        State University, May 1977.
6- 9.   C. G. Frye, J. Chem. Eng. Data, 7, 592 (1962).
6-10.   T. P. Wilson, E. G. Caflisch and G. H. Hurley, J. Phys.
        Chem., 62, 1059 (1958).
6-11.   L. D. Rollmann, J. Cat, 46, 243 (1977).
6-12.   R. J. Hengstebeck, "Petroleum Processing", McGraw-Hill
        Book Co., Inc., New York, 1959, p. 279.
6-13.   A. Szekely, Acta Chim. Hung., 5, 317 (1955).
6-14.   G. Rabo and A. Szekely, ibid., 5, 453 (1955).
6-15.   D. D. Whitehurst, T. O. Mitchell, M. Farcasiu and J. J.
        Dickert, Jr., "The Nature and Origin of Asphaltenes in
        Processed Coals", Final Report Under Project RP-410-1,
        March 1977-January 1979, in press.
6-16.   D. D. Whitehurst, T. O. Mitchell, M. Farcasiu, and J. J.
        Dickert, Jr., "Exploratory Studies in Catalytic Coal
        Liquefaction", Final Report Under EPRI Project 779-18,
        May 1979, in press.
6-17.   W. Radmacher and P. Mohrhauer, Brennst. Chemie, 36, 236
        (1955).
6-18.   M. Bishop and O. L. Ward, Fuel, London, 37, 191 (1958).
6-19.   P. Painter, The Pennsylvania State University, personal
        communication, 1977.
6-20.   R. A. Meyers, "Coal Desulfurization", Marcel Dekker, Inc.,
        New York, 1977.
6-21.   P. H. Given, P. L. Walker, W. Spackman, A. Davis, and H. L.
        Lovell, "The Relation of Coal Characteristics to Coal
        Liquefaction Behavior", Report No. 4, Special Report from
        the Pennsylvania State University to NSF, December 1, 1975.
6-22.   C. B. Marson, Gas Journal, 171, 39 (1925).

Chapter 7

EFFECTS OF PROCESS VARIABLES (TIME, TEMPERATURE,

AND HYDROGEN GAS PRESSURE) IN SRC PRODUCTION FROM COAL

I.   INTRODUCTION

For an optimal SRC production, several critical process factors must be balanced. The optimal product yield and quality can
be achieved by controlling the time of conversion, the temperature,
the properties of the solvent, and the pressure of hydrogen gas.
Economics dictate that the conversion times be short so that
equipment can be small. Temperatures should not be too high because of materials expense and potential excess gas and coke production. Pressurizing with hydrogen can be advantageous to a
point, since increasing the reaction rates lowers the needed equipment size; however, high pressures again raise expenses because
heavier equipment must be used.
In this chapter we describe how these various parameters can
affect the overall course of the conversion of coal to SRC. The
initial discussions will concentrate on how coal liquefaction
features such as conversion rate, product composition, hydrogen
consumption, etc. are affected by increasing the reaction time
and hydrogen pressure at relatively mild temperatures ($800°F$).
These features will then be contrasted with those which result
when the conversion temperature is raised.

II.   CONVERSION AT $800°F$

A.   Intrinsic Solubility of Coal Components

Assessing the conversion of coals to soluble products is
complicated because nearly all coals intrinsically contain some

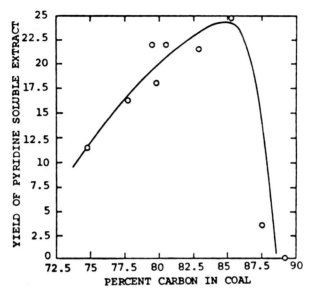

FIGURE 7-1. Yield of Extract (Percent Weight Recovered) vs. Rank (% MAF Carbon)

constituents which are soluble in common organic solvents. Dryden (7-1) and others have described rather extensively the amount and chemical nature of the soluble constituents of coal. In this book, "conversion" is defined as the yield of gases plus pyridine-soluble materials produced on subjecting coal to various treatments.

Jenkins has shown that pyridine solubilities as high as 40% can be found for some bituminous coals (7-2). The yield of pyridine-soluble materials passes through a maximum as one proceeds from low to high rank coals, as shown in Figure 7-1 for coals investigated in MRDC work.

The composition of this intrinsically extractable material is very similar to the whole coal. The oxygen contents of low rank coals are somewhat lower in extracts than in whole coals but for most bituminous coals the extracts' and the coals' oxygen contents were similar (Figure 7-2). The organic sulfur contents vary as a function of geological and geographical origin of the coal, rather than its rank, but the sulfur contents of the extracts correlate with the organic sulfur of the coal (see Figure 7-3). The H/C mole ratios of extracts increase with decreasing rank and no doubt reflect the lower aromaticity of low rank coals (Figure 7-4).

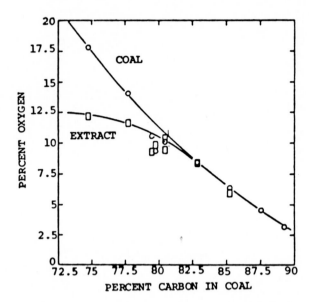

FIGURE 7-2.   Oxygen Content of Extract and Coal vs. Rank

    Detailed SESC characterization of extracts showed that the
bituminous coals, which have large contents of soluble components
(7-3) have high proportions of the most polar fractions (asphaltols). Lower rank (Western) coals have lower overall solubilities
and the extracts contain less asphaltols than those from higher
rank coals.
    The yields and compositions of the extracts were used as a
basis for evaluating the changes which occur in coal on pretreatment or reaction at elevated temperatures.
    The presence of intrinsically pyridine-soluble material in
coals raises the following important questions about the earliest
stages of coal conversion: Does the initial product contain most
of the intrinsically soluble material? How does the presence of
this material (and its removal) affect the kinetics and mechanisms
of dissolution of the insoluble portion? Does its presence or
absence affect chemical reactions of the remainder, especially
with regard to SESC fraction and residue distribution? What is
the effect on hydrogen consumption?

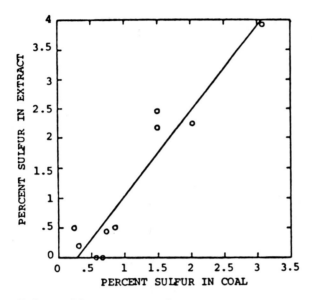

FIGURE 7-3. *Sulfur Content of Extract vs. Sulfur Content of Coal*

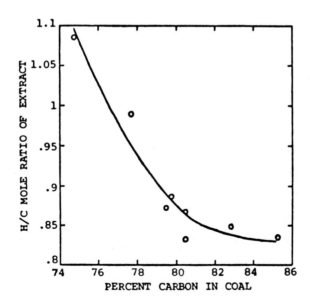

FIGURE 7-4. *H/C Mole Ratio of Extract vs. Rank*

In an effort to answer such questions, the pyridine-soluble
and insoluble portions of West Kentucky 9,14 coal were separated
by extraction and each was reacted with a synthetic recycle solvent (SS*43) at 800°F and 1000 psi H_2. These results were compared with those of whole coal reacted under the same conditions
(7-5).

The pyridine-insoluble portion of the coal was found to dissolve at essentially the same rate as the whole coal, and these
results indicate that the presence of extractable components in
coal is not necessary for rapid dissolution of coal. Surprisingly,
the pyridine-soluble portion produced 13.6% insoluble residua
after only 4.5 minutes reaction. Although the production of H_2O
H_2S, and CH_4 was the same in conversions of extract and residue,
the pyridine-soluble feed produced 4.71% CO and only 0.83% CO_2
while the pyridine-insoluble portion of the coal produced 1.35%
CO_2 and no CO. This suggests that the form of the oxygen in the
two portions may be different. The whole coal produced more CO_2
than CO. Interestingly, the yield of light hydrocarbon gases was
greater from the insoluble portion. The yield of SRC was essentially the same.

Perhaps most important was that the hydrogen removed from the
solvent was also essentially the same in the extract and residue.
This is shown in Figure 7-5 where the calculated hydrogen consumption from solvent is plotted vs. time for the 2 runs. Data for the

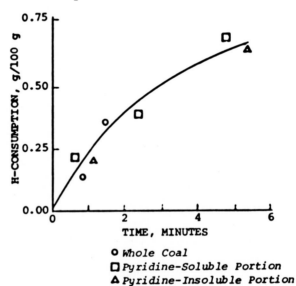

○ *Whole Coal*
□ *Pyridine-Soluble Portion*
△ *Pyridine-Insoluble Portion*

FIGURE 7-5.   *Hydrogen Consumption from Solvent vs. Time*
for West Kentucky 9,14 Coal

whole West Kentucky 9,14 coal are also included. In the first
five minutes of reaction, all three feeds behaved in the same
manner, thus the process that renders the coal soluble does not
require a detectable excess of hydrogen over that consumed by the
components in solution. Either dissolution of this bituminous
coal requires little hydrogen or it involves the same kinds of
reactions that the dissolved species undergo. These could be
reactions of the skeleton or of the functional groups in the
system. (However, a difference in the mode of oxygen elimination
was noted above.)

Elemental analyses of the feeds and products, SESC analyses,
and ^{13}C-NMR data showed again a striking similarity of results in
the two runs. The two fractions when separated appear to convert
in an essentially identical manner. Any slight differences in the
aromatic H content of the two SRCs reflect the differences in the
aromatic C content of the starting feeds. These results indicate
that the portion of coal which can be directly extracted is chemically and structurally very similar to the non-extractable material
for West Kentucky 9,14 coal, and its presence or absence does not
seriously affect the chemistry or kinetics of coal dissolution
(7-5).

This may not be a general phenomenon, however. For example,
Western coals (Wyodak-Anderson) generally have extracts much
richer in hydrogen content (7%) than the residues (about 5%), and
might behave differently on conversion than bituminous coals (7-4).

B. Rate of Coal Dissolution

It has been known for a number of years that some coals can
be solubilized at very short times (7-3 through 7-8). The results
described in this section were generated in two ways. In most
experiments, short runs were initiated by coal injection to
achieve rapid heatup and terminated by rapid cooling. In some
instances, samples were withdrawn at specified times during longer
runs through a filter and the SRC was isolated.

Standard conditions were 800°F, 1000 psi H$_2$, and synthetic
recycle solvents containing 43% tetralin.

Using apparatus in which the time and temperature could be
precisely measured, the rates at which coals became soluble were
found to vary dramatically. A portion of all coals solubilizes
extremely rapidly; this is paralleled by the kinetics of oxygen
conversion as shown in Figures 7-6 and 7-7 which show very rapid
initial oxygen conversion (7-4).

For bituminous coals, over 80% of the coal can be solubilized
in about 3 minutes. The quantity of easily dissolved material in
subbituminous coals is somewhat less (60%) and is produced more

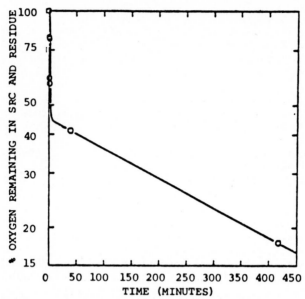

FIGURE 7-6. *Kinetics of Oxygen Removal from West Kentucky 9,14 Coal*

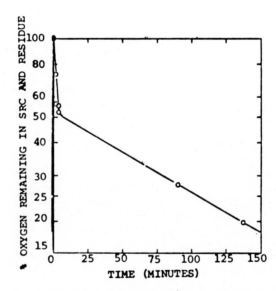

FIGURE 7-7. *Kinetics of Oxygen Removal from Wyodak-Anderson Coal*

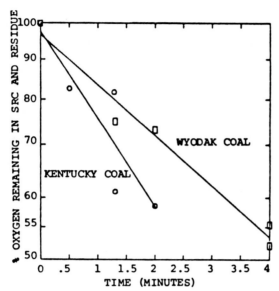

FIGURE 7-8. *Kinetics of Rapid O Loss of West Kentucky 9,14
and Wyodak-Anderson Coals*

slowly. In both cases the rate of rapid dissolution appears to
be approximately first order in coal as shown in Figure 7-8.

(In solvents of low hydroaromatics content the rates of coal
dissolution are slower. Explanations for this are presented in
detail in the chapter on solvent chemistry.)

Figure 7-9 compares the conversions at 3 minutes for a wide
range of coals. The maximum at 85% MAF carbon in the coal
corresponds roughly to the maximum yield of intrinsically soluble
material (compared with Figure 7-1).

Coals of this rank are also known to have the greatest
plasticity and swelling properties (7-10). Thus rapid dissolution
may depend on the physical as well as chemical properties of the
coal. Chemical factors leading to rapid solubility of coals are
discussed in detail in chapters on coal rank effects on liquefaction and on model compound conversions. We shall merely state
here that rapid conversion appears to be associated with the
presence of a sufficient number of easily broken bonds in the coal
structure. These correspond to benzylic ethers, ethylene linkages
between aromatic rings and certain other alkylsubstituted
aromatics. The low reactivity of high rank coals (>89% C) is
believed to be due to the presence of more highly condensed
structures, especially condensed aromatic rings (7-3).

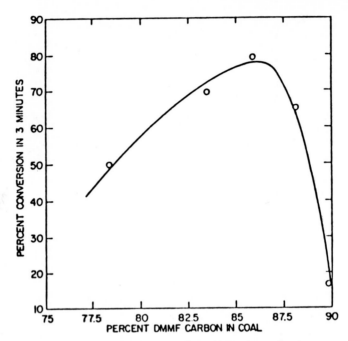

FIGURE 7-9. *Conversion of Coals of Various Ranks*

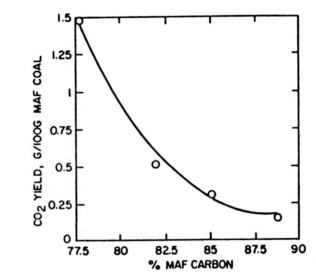

FIGURE 7-10. *CO_2 Yields From Coals of Various Ranks*

C. Composition of the Initial Products

The compositions of the initial products of coal solubilization closely resemble the compositions of the parent coals. The
carbon-type distributions are essentially indentical (as shown
in Table 7-1). The oxygen contents of short-contact-time SRCs
are very similar to the bituminous parent coals, but are significantly lower for subbituminous coals.

The chemical and physical properties of the SRCs depend on the
amount and type of oxygen present. At short times, about 40-50%
of the initial oxygen is easily removed from coal. Sulfur losses
parallel oxygen losses (7-5).

Several reactions lead to rapid oxygen loss in the SRCs. The
phenolic content of SRCs has been shown to closely resemble the
phenolic content of coals of the same carbon content at high conversions (>60%) (7-5). At lower conversions the soluble coal
products contain significantly fewer phenols than the parent coal.
This may indicate that the most easily dissolved components of
coal have less functional groups/average molecule. Loss of some
functionalities are, however, as rapid as coal dissolution. Dehydroxylation of polyaromatic phenols is fairly rapid and benzylic
oxygenates such as alcohols or esters are rapidly decomposed under
liquefaction conditions (7-3).

The evolution of CO_2 from coal or coal products is very fast
but the amount of CO_2 evolved is a function of the rank of the
coal. Lower ranks produce more CO_2 (see Figure 7-10).

The average molecular weights of the early products of coal
dissolution are rather high (2000) or the molecules are highly
associated because of their high functionality. After 5 minutes
of conversion, the molecular weights become lower (300-1000).
Further reduction in molecular weight requires much longer reaction times.

Under conditions where the conversions are reasonably high,
the initial SRC is found to be predominantly composed of
asphaltenes and asphaltols. The major constituent varies with
rank. Bituminous coals, which produce the largest yields at
short times, are rich in asphaltols. In some cases as much as
85% of the product can be benzene insoluble (asphaltols) (7-3).

Initial products from subbituminous coals contain relatively
more asphaltenes than those from bituminous coals, (Figure 7-11),
which is rather surprising as the oxygen content of short-
contact-time SRCs continually increases with decreasing rank and
is proportional to the oxygen content of the coal (see Figure 7-
12). These data indicate that ethers are somewhat more prevalent
in low rank coals than in high rank coals. The sulfur content of
short-contact-time SRCs is also proportional to the sulfur content of the parent coals as shown in Figure 7-13.

TABLE 7-1. *Aromatic Carbon Content of Coals and Short Contact Time SRC*

Sample	Temperature	Time (min)	% Conversion	% Carbon (CP-^{13}C-NMR) Aromatic	Alkoxy	Aliphatic
Wyodak-Anderson Coal	–	–	–	69	6	25
Wyodak-Anderson SRC	860	3.0	67.0	70	3	27
Illinois #6 (Burning Star) Coal	–	–	–	67	–	33
Illinois #6 (Burning Star) SRC	792	0.75	77.1	72	–	28
Illinois #6 (Burning Star) SRC	860	1.25	88.6	74	–	25
Illinois #6 (Monterey) Coal	–	–	–	69-71	–	31
Illinois #6 (Monterey) SRC	800	2.0	81.2	70	–	30
Illinois #6 (Monterey) SRC	800	4.0	88.2	76	–	24
West Kentucky 9,14 Coal	–	–	–	66	–	34
West Kentucky 9,14 SRC	800	0.5	50	60	–	40

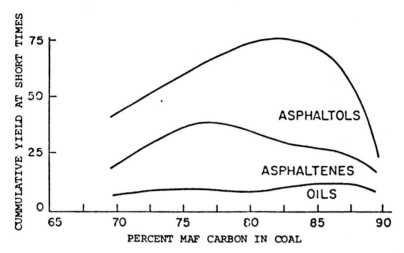

FIGURE 7-11.   Composition of Various Short Contact Time SRCs

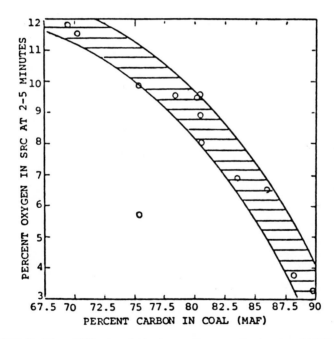

FIGURE 7-12.   Effect of Rank on Oxygen Content of SCT-SRC

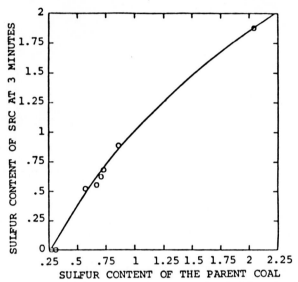

FIGURE 7-13.   *Relationship Between the Sulfur Contents of*
Coals and Short Contact Time SRCs

D.   Hydrogen Consumption

As coal dissolves, hydrogen must be consumed. Free radicals
are formed from the thermal disruption of weak bonds in the coal
structure; for solubility to be achieved, these radicals must be
capped by hydrogen or they will recombine. This hydrogen can
come from a variety of sources: the solvent, gaseous H_2, or from
the coal itself.

The most efficient source is hydroaromatics in the solvent but
if such materials are limited in concentration, hydrogen gas or
coal become the dominant sources. This is illustrated in Figure
7-14 for a series of coals reacted in the presence of a solvent
containing only 8.5% tetralin. The figure clearly shows that
even at short times hydrogen gas can be the dominant source of
hydrogen for low rank coals where the demand for hydrogen is
largest (7-3).

E.   Sensitivity to H-Donors, H_2, and H-Shuttlers

The rate of coal dissolution is proportional to the concentration of hydroaromatics in synthetic recycle solvents (7-3).
A clear example of this can be found in the conversion of
Illinois #6 (Burning Star) coal at 3 minutes in a series of solvents with varying tetralin concentrations (see Figure 7-15).

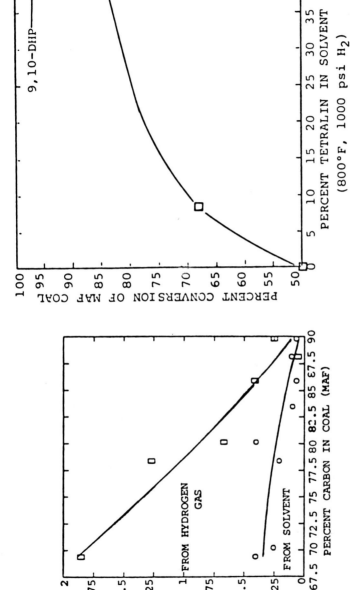

FIGURE 7-15. Conversion of Bituminous Coal, Illinois #6 (Burning Star) in 3 Minutes at Various Donor Levels

FIGURE 7-14. Hydrogen Consumption at Short Time for Coals of Various Ranks

This simple relationship is complicated somewhat by hydrogen donation from other sources such as H_2 gas or the coal itself. It has been our experience that low rank coals can give increased yields at short times by application of H_2 pressure. For higher rank coals (bituminous) the effects are very small (7-3,7-4).

It is suggested that the bituminous coals are efficient sources of hydrogen because a high proportion of the mass is plastic or mobile at liquefaction temperatures.

The donation of hydrogen from hydroaromatic structures in coal can be assisted by certain highly condensed aromatic molecules in solvents. Such molecules are not net donors of hydrogen but can rapidly equilibrate with hydroaromatics in the coal and can thus "shuttle" hydrogen from one region of the coal to another. This topic is discussed in detail in the chapter on solvent chemistry, but as an example, Figure 7-16 shows a group of solvents of limited H-donor capacity where the amount of coal becoming soluble in 4 minutes is proportional to the concentration of poly-condensed aromatic compounds in the solvent (7-5). It is note-worthy that a good shuttling solvent can even induce higher solubility than a solvent containing 40% tetralin (SS in the figure).

In summary, large portions of all coals rapidly dissolve at short reaction times. Bituminous coals give the highest yields and require little hydrogen, but the presence of either good hydrogen donors or hydrogen shuttlers is necessary for high

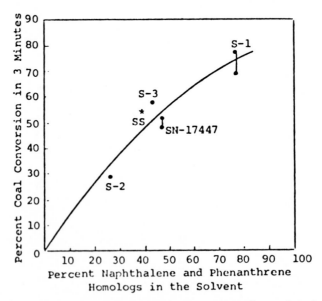

FIGURE 7-16.   *Coal Conversions as a Function of Poly-aromatic Solvent Compounds*

conversions at short times. Subbituminous or lower rank coals can give high yields of soluble material but at a slower rate.

F.   Progress of Reactions With Extended Times

    1. Compositional Changes

    Though SRCs obtained from different coals at short reaction times retain most of their coals' chemical structures, these distinctive factors are lost as the reaction time is extended (7-5). All coals tend toward a common composition, which is dictated predominantly by the reaction conditions and, to a lesser extent, by the composition of the solvent used in the reaction. SRCs from different coals all tend toward a common aromatic carbon content (at 800°F, SS*43) of 75 percent. Residues tend toward 85% aromatic carbon.
    Figure 7-17 shows how the aromatic contents of SRCs from two different coals are similar at longer times.

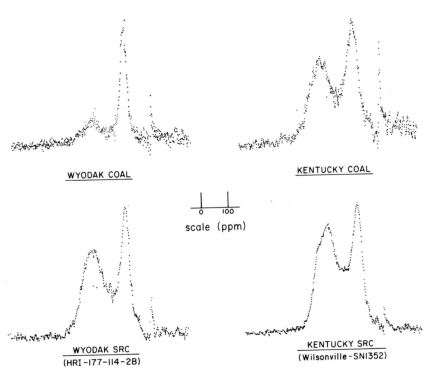

FIGURE 7-17. CP-^{13}C-NMR Spectra of Coals and SRCs

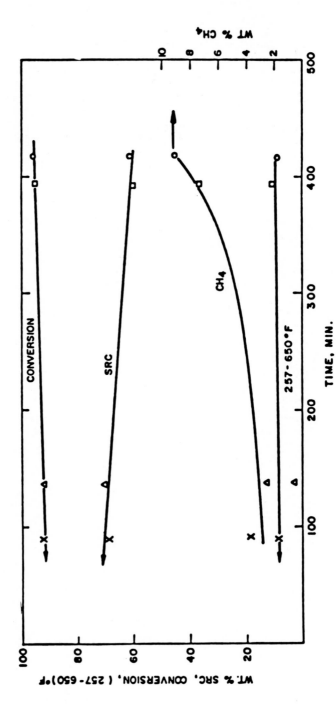

FIGURE 7-18. *Product Distributions at Low Temperatures and Long Times.*

In the SRC process, the H/C mole ratio of the SRCs continually declines as the reaction time is increased. This decline is due to the hydrogen demand which is generally greater than can be supplied rapidly by the solvent or gaseous hydrogen. Thus coal or coal products contribute substantially to hydrogen donation and become depleted. Also, at liquefaction temperatures thermodynamics favors dehydrogenation of hydroaromatic structures in coal (7-3).

The gross product compositions are similar for a variety of coals at times greater than 90 minutes. As seen in Figure 7-18, coals tend to yield common product distributions when reacted with solvents containing high concentrations of H-donors (43% tetralin). The elemental analyses of long contact time SRCs also become similar (7-3, 7-4, 7-5).

The transition between long and short contact times follows a similar course for all coals. Asphaltols, which are the predominant products of coals, are converted simultaneously to oils, asphaltenes, and insoluble residues (char) (7-11).

Thus, oils and asphaltenes are formed simultaneously rather than sequentially. Although there is some conversion of asphaltenes to oils, with bituminous coals the end products of the SRC process are predominantly heterocyclic aromatics (7-3). Sub-bituminous coals, on the other hand, contain more oxygen which is largely found in ether structures at short reaction times. On extended reaction times, a large portion of the etheric bonds break to produce phenolic groups. Monophenols on single aromatic rings appear to be the predominant end products of SRC liquefaction of these lower rank coals.

2. Hydrogen Consumption

Although the amount of hydrogen required to solubilize coals at short times is small, the rate of demand for hydrogen is greatest during the first few minutes of coal conversion processes. However, there are measurable rates of consumption even after extended periods of time. In the initial phase, hydrogen is needed to cap thermally produced radicals, and to aid in eliminating reactive sulfur species and oxygen from polyaromatic ring phenols.

As the conversion proceeds, hydrogen is consumed for heteroatom removal, but at much slower rates (refer to Figure 7-6 and 7-7). The proportion of the hydrogen consumed for heteroatom rejection alone becomes a small part of the total; light hydrocarbon gases and solvent-range products account for the majority. This point is illustrated for West Kentucky 9,14 coal in Figures 7-19 and 7-20. Hydrogen consumption is nearly stoichiometric

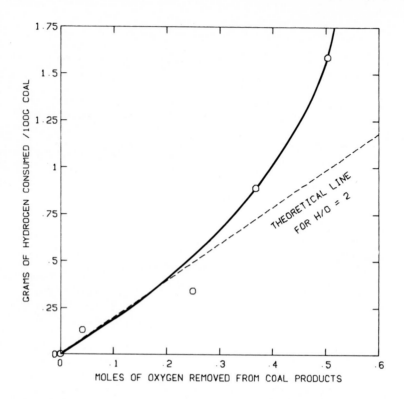

FIGURE 7-19. Sensitivity of H Consumption to Oxygen Removal

with oxygen removal in the early stages but is more closely related to light gas and solvent production in the later stages.

Figure 7-20 also emphasizes the fact that light hydrocarbon
gas formation always accompanies solvent production in thermal
processes; thus, there is room for improvement in the efficient
utilization of hydrogen by changing this selectivity.

3. Sensitivity to H-Donors, H_2 and H-Shuttlers

In thermal coal liquefaction processes, the total hydrogen
consumption is generally less than 5% of the amount of coal feed.
Solvents rich in hydrogen donors can provide the required hydrogen without the aid of gaseous hydrogen. It has been reported
that even in the conversion of coal to distillates, hydrogen gas
over-pressure provides no additional conversion when solvents of
high hydroaromtic content are used (7-12).

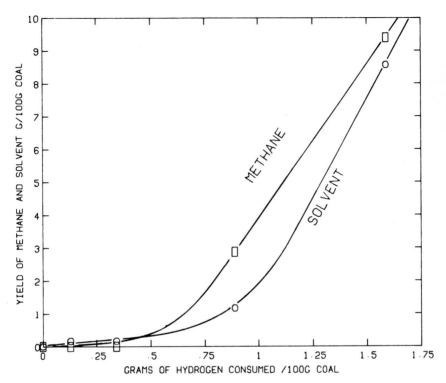

FIGURE 7-20. Requirement of Hydrogen For Gas & Solvent Production

However, in most of the processes being developed, solvents of such high hydroaromatic contents are not practical even when separate external hydrogenation of the solvent is employed. In the Exxon Donor Solvent (EDS) process, about one-half of the hydrogen needed to produce the distillate product is provided directly by hydrogen gas (7-13). On reaction with coal the solvent's hydrogen content is reduced by only 1-2 g/100g solvent before the spent solvent is regenerated. In other coal liquefaction processes, hydrogen gas is required to maintain the low level of hydrogen donors present in the recycle solvent. It is believed that coal minerals, especially pyrite, are necessary to catalyze solvent rehydrogenation by hydrogen gas. Without such catalysts, conversions are limited by the amount of hydrogen contained in the solvent or available in hydroaromatic structures in the coal itself.

To illustrate this point, we shall consider a series of experiments in which Illinois #6 (Monterey) coal was reacted for 90 minutes in the presence of 1000 psi H_2 and synthetic recycle solvents of varying tetralin contents. In this series 2.5 g of hydrogen per 100 g coal were consumed in each case in rejecting heteroatoms and forming hydrocarbon gases and distillates. The amount of hydrogen provided by each of the various sources is shown below (7-3):

SOURCES OF HYDROGEN DURING COAL LIQUEFACTION

% Tetralin in the Solvent	40	8.5	4	0
% Of Total H Donated by:				
Hydroaromatics in Solvent	80	32	19	0
Hydrogen Gas	0	31	43	57
Coal	20	37	38	43

It is clear that as hydroaromatics are depleted from the solvent, hydrogen gas and coal make up the deficiency.

For a Western subbituminous coal (Wyodak-Anderson), hydrogen gas plays a less important role, providing only 30% of the required hydrogen in the solvent containing no hydroaromatics. In that case coal provided the majority of the needed hydrogen, as it is suspected that the mineral matter (which contains no pyrites) could not catalyze utilization of H_2 gas.

Some confirmation for this supposition has been found (7-3) by selective removal of the pyrite in a bituminous coal and reaction of the demineralized and parent coals in solvents which contained no hydroaromatics. In those experiments the hydrogen gas consumption was 2 g/100 g coal for the parent coal and only 1 g/100 g coal for the demineralized coal.

The consequence of hydrogen donation by coal or coal products is lower quality SRC, lower conversion and often char formation. Detailed discussions on the formation of char in H-donor limited solvents is provided in the chapter on the mechanisms of char formation.

Hydrogen shuttlers can also play a role in coal conversions at long times. The effect is to aid in rapidly depleting the hydrogen content of the coal products. If this depletion is excessive, char formation will result (7-3).

III. CONVERSIONS AT HIGH TEMPERATURES (820-880°F)

A. Effect of Rates of Conversion and Product Compositions

1. Short Contact Time

Upon increasing the temperature of reaction during coal liquefaction all reaction rates increase. This includes rates of coal dissolution, heteroatom rejection, hydrogen consumption, gas formation, and charring (7-3). The fact that the rate of char formation increases often confuses the interpretation of the effect of temperature on coal conversion, as both char and unconverted coal are insoluble.

The balance between achieving desired reactions while avoiding char formation is always critical at elevated temperatures. It has been reported (7-7) that for a low rank coal (Akiva) raising the temperature above 800°F lowered the degree of coal conversion at 2 min. Secondary condensation reactions of initially dissolved coal producing insoluble products was suggested as an explanation for this result.

One major incentive for using elevated temperatures is to decrease the size of the reactors needed for coal liquefaction by increasing the rates. In examining the behavior of a number of coals it was concluded (7-3) that for bituminous coals little benefit is achieved on raising the temperatures of conversion to 845°F if a short contact time product is desired. Figures 7-21, 7-22, and 7-23 show that 80% conversion is achieved in less than 3 min. for three bituminous coals. Conversions at 845°F do give faster rates but in the time spans which are practical in continuous flow reactors, differentiating between 1 and 3 min. residence time is difficult. Only by raising the temperatures to very high levels (900°F) can significant changes in the composition of the SRC be achieved at times less than 5 min.

With subbituminous coals, or lower ranks, the rates of coal dissolution are considerable slower than for bituminous coals (7-3). Thus raising the temperature may be desirable for dissolving coal in less than 5 minutes. With one subbituminous coal (Wyodak-Anderson) increasing the temperature to 820, 840, 850 and 860°F gave increasing conversions at short times but at 850-860°F the results were found to be highly erratic as shown in Figure 7-24. Sensitivity of conversions were attributed to minor variations in the coal composition (minerals and aromaticity) and to the difficulty in reproducing precise time temperature profiles at high temperatures and short times. No major char formation was noted for this coal at temperatures up to 860°F; even with high scatter in the conversion data, the selectivity of conversion vs. hydrogen consumption was nearly the same for all the reactions (7-3).

    2.   Long Contact Time

    The net effect of increasing the temperature of reaction in long contact time coal conversion is to decrease the SRC yield and increase the yield of light hydrocarbon gases (see Tables 7-2 and 7-3 and Figures 7-25 and 7-26). This is true either with or without hydrogen donors in the solvent. Heteroatom rejection is also accelerated and provides SRCs of

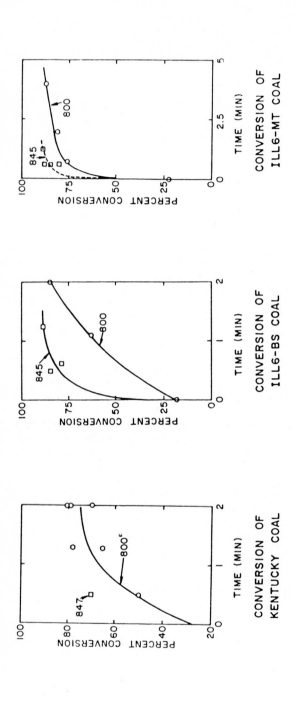

CONVERSION OF
KENTUCKY COAL

CONVERSION OF
ILL6-BS COAL

CONVERSION OF
ILL6-MT COAL

FIGURE 7-21. Conversion
of W. Kentucky 9,14 Coal.

FIGURE 7-22. Conversion
of Ill. #6 (Burning Star)
Coal.

FIGURE 7-23. Conversion
of Ill. #6 (Monterey)Coal.

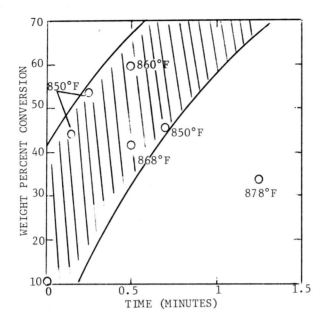

FIGURE 7-24.   Conversion of Wyodak Coal vs. Time

lower sulfur content, but the H/C mole ratios of the products are
lower as well.   Char formation is also accelerated at higher
temperatures.   The SRCs of high temperature conversions also contain lower concentrations of highly polar fractions and are therefore more soluble in hydrocarbons (7-3).

B.   Effect on Hydrogen Consumption

      As stated above, raising reaction temperatures accelerates all
reactions, including those which consume hydrogen.   Increased
hydrogen consumption, though necessary for heteroatom removal, can
be excessive at high temperatures in that much of the increase is
due to the formation of light hydrocarbon gases (see Tables 7-2
and 7-3).   Often none of the increased hydrogen requirement is
reflected in the SRC quality, as the SRCs of high temperature conversions generally have low H/C mole ratios (7-3,7-5).
      These data indicate that raising the temperature of reaction
solely to increase the rates of heteroatom removal may not be
desirable from the standpoint of efficient hydrogen utilization.
Another disadvantage of the production of gaseous hydrocarbons is
that such gases dilute the hydrogen recycle stream.   Highly dilute

D. Duayne Whitehurst *et al.*

TABLE 7-2. Effect of Temperature and Donor Concentration
on Monterey Coal Conversion
(90 minutes, 1500 psi initial H_2 pressure, solvent/coal ~6)

Solvent	$SS*43^a$	$SS*43^a$	$SS*0^b$	$SS*0^b$
Temperature, ^OF	800.00	858.00	860.00	800.00
MAF Conversion, wt%	91.96	78.06	54.29	70.68
H_2S	1.28	2.11		1.17
Water	1.86	5.66	5.30	5.99
CO	0.35	0.99	0.25	0.44
CO_2	1.94	2.87	3.58	2.94
C_1	3.12	7.01	5.14	2.77
C_2-C_5	3.58	7.20	4.19	4.59
(C_6-470^OF)	3.72	1.37	0.05	0.43
$(470-650^OF)$	4.68	3.52	1.96	0.38
SRC	69.62	46.50	39.82	54.20
MAF Residue	8.04	21.96	45.71	29.32
Balance	98.19	99.20	106.00	102.23
H_2 Consumption (solv.)	1.97	3.87		0.07
H_2 Consumption (gas)	0.03	0.14	1.08	1.34
H_2 Consumption (total)	1.94	4.01	1.08	1.41

[a] 43 Percent tetralin in solvent.

[b] 0 Percent tetralin in solvent.

TABLE 7-3. Effects of Temperature and Donor Concentration
on Wyodak Coal Conversion
(137.5 min., 1500 psi initial H_2 pressure, 43% tetralin in solvent)

	800.00	820.00	840.00
Temperature, $°F$	800.00	820.00	840.00
MAF Conversion, wt%	91.50	79.58	81.55
Solvent/Coal	6.17	3.97	4.21
H_2S		0.13	
Water	4.59	0.02	0.16
CO	1.28	1.18	2.09
CO_2	6.04	5.48	10.12
C_1	2.50	1.87	4.10
C_2-C_5	4.53	3.57	7.37
$(C_6-470°F)$	0.14	2.07	1.46
$(470-650°F)$	2.38	1.79	1.30
SRC	70.00	63.46	54.95
MAF Residue	8.50	16.33	17.42
H_2 Consumption (solv.)	1.80	3.05	3.39
H_2 Consumption (gas)	0.06	0.83	0.75
H_2 Consumption (total)	2.58	3.88	4.14

Note: All balances forced to 100 percent.

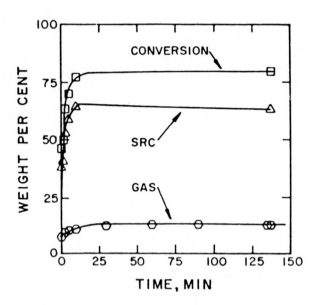

FIGURE 7-25. *Wyodak Conversion at 820°F.*

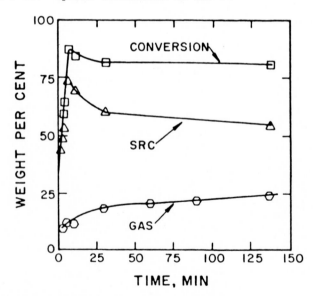

FIGURE 7-26. *Wyodak Conversion at 840°F.*

gas streams cannot be cleaned and much hydrogen is lost in waste streams.

C. Sensitivity to H-Donors and H_2

The need for hydroaromatic constituents in recycle solvents is critical when the temperature of coal conversion is raised. With increasing temperature a larger, faster demand for hydrogen is created and coal products become more prone to char formation through hydrogen abstraction. Table 7-2 shows a clear trend of lower "conversion" at 860°F than at 800°F. With no hydroaromatics in the solvent, the yields of SRC are lower and yields of light hydrocarbon gases are higher. Solvent yields are not appreciably increased on raising the temperature at long reaction times.

Because of the larger demand for hydrogen at high temperatures, the significance of H_2 gas pressure is accentuated. Thermo-dynamics also become less favorable for retention of cyclic aliphatic structures and high hydrogen pressures are needed to counter the shift in equilibrium towards aromatics.

IV. SUMMARY

The inherently pyridine-soluble and pyridine-insoluble por-tions of bituminous coals appear to have similar liquefaction reactivities and product distribution. Coals can be very rapidly solubilized; the compositions of the initial products closely resemble the parent coals.

The hydrogen demand created as coals solubilize will be met by the solvent, gaseous H_2, or the coal itself. Solvent hydro-aromatics are the best hydrogen source and their concentration and donor ability control the rate of appearance of soluble products. H_2 gas pressure increases the yields, especially for lower rank coals in poor donor solvents. This is catalyzed by coal mineral matter. Supply of hydrogen by the coal itself can result in charring and poor quality SRCs. At short times the rate of hydrogen demand is high but the amounts required may be low, especially for bituminous coals. At longer times the total amounts required may become large.

Initial products vary with rank but are always rich in asphaltols. At longer times asphaltols are converted simulta-neously to oils and asphaltenes, and SRC compositions from different coals become very similar.

Raising the reaction temperature speeds up all reaction rates and makes the progress of conversion much more sensitive to the other reaction variables -- such as H_2 overpressure and solvent hydroaromatic content. For bituminous coals little advantage was found. Conversion of soluble coal products to char can be a serious problem at high temperature. Although the rejection of heteroatoms is more favorable at high temperature, so is the rate of hydrocarbon gas formation. The end result is that hydrogen is not as efficiently utilized.

REFERENCES

7- 1.   I. G. C. Dryden, Nature, <u>163</u>, 141 (1949).
W. Francis, <u>Coal</u>, Edward Arnold, Ltd., London, 1961, pp 458-462.
7- 2.   R. G. Jenkins, to be published.
7- 3.   D. D. Whitehurst, T. O. Mitchell, M. Farcasiu and J. J. Dickert, Jr., "The Nature and Origin of Asphaltenes in Processed Coals", Final Report Under Project RP-410-1, March 1977-January 1979, in preparation.
7- 4.   D. D. Whitehurst, M. Farcasiu and T. O. Mitchell, "The Nature and Origin of Asphaltenes in Processed Coals", EPRI Report AF-252, First Annual Report Under Project RP-410, February 1976.
7- 5.   D. D. Whitehurst, M. Farcasiu, T. O. Mitchell and J. J. Dickert, Jr., "The Nature and Origin of Asphaltenes in Processed Coals", EPRI Report AF-480, Second Annual Report Under Project RP-410-1, July 1977.
7- 6.   G. P. Curran, R. T. Struck and E. Gorin, <u>I&EC Proc. Design. and Develop.</u>, <u>6</u>, 166 (1973).
7- 7.   M. Morita and K. Herosawa, Nenryo Kyokai, <u>Shi</u>, <u>53</u>, 263 (1971).
7- 8.   R. C. Neavel, Proc. Symp. on Agglomeration and Conversion of Coal, Morgantown, West Virginia, 1975 (April 1976).
7- 9.   W. Spackman et al., "The Characteristics of American Coals in Relation to Their Conversion into Clean Energy Fuels", Quarterly Technical Progress Report October-December 1977 (March 1978), FE-2030-10, to U. S. D. O. E. under contract EX-76-C-01-2030.
7-10.   Y. Sanada and H. Honda, <u>FUEL</u>, <u>45</u>, 295 (1966).
7-11.   M. Farcasiu, T. O. Mitchell and D. D. Whitehurst, Chemtech, <u>7</u>, 690 (1977).
7-12.   L. E. Furlong, E. Effron, L. W. Vernon and E. L. Wilson, "Coal Liquefaction by the Exxon Donor Solvent Process", presented at AICHE National Meeting, Los Angeles, CA., Nov. 18, 1975.
7-13.   EDS Coal Liquefaction Progress Development Phases IIIB/IV, Annual Technical Progress Report for the Period July 1, 1977 to June 30, 1978, Prepared for U. S. D. O. E. Under Cooperative Agreement EF-77-A-01-2893 (September 1978).

Chapter 8

THE FORMATION OF CHAR DURING COAL LIQUEFACTION

I. GENERAL COMMENTS

Residue or char is defined in this chapter as the pyridine-insoluble product obtained from heating coal or coal liquids. A portion of these residues may have been pyridine-soluble initially, but they became pyridine-insoluble due to regressive reactions. In charring* experiments using whole coal, the residue also includes unreacted coal. Fortunately, optical microscopy can distinguish between the unreacted coal and the char product formed. The optical analyses presented were obtained through a joint research program with the College of Earth and Mineral Sciences, The Pennsylvania State University (Penn State) (8-3).

The optical characteristics of insoluble carbonaceous residues or cokes are described in detail in several excellent reviews (8-4 - 8-7). A brief summary of various cokes is given in Table 8-1.

Production of these residues can be desirable or disastrous, depending on the particular situation in which they form. "Coking" coals with strong physical properties are used extensively in the manufacture of metallurgical coke. Electrodes for aluminum and steel manufacture are made from coal tar pitches or petroleum coke which yield "graphitizable" (highly ordered) carbon. Glassy carbon is used to fabricate many high-strength carbon articles. In coal liquefaction processes, however, solids formation must be prevented or at least controlled, or the process transfer lines and/or reactors will rapidly plug.

Researchers have extensively studied the reaction mechanisms, critical conditions and sequential formation of anisotropic and isotropic cokes (8-4 - 8-10). The classic work of J. D. Brooks and G. H. Taylor (8-4) first described mesophase carbon; in

*The terms "coking", "charring" and "carbonizing" are used interchangeably in this text.

TABLE 8-1.   *Types of Organic Pyridine-Insoluble Residues*

Unreacted Coal -

   Altered Coal - Particles having the shape of the original
                    coal but exhibiting much higher reflectance.

     Vitroplast - A plastic or once-plastic and optically iso-
                    tropic (nonordered) degradation product of
                    vitrinite (also from huminite and semi-
                    fusinite).

Granular Residue - Particles less than 1 μm which are not identi-
                     fiable as individual components; includes both
                     maceral and mineral-derived materials.

    Agglomerate - Large particles composed of agglomerated
                    smaller particles of coal, altered coal,
                    and/or mineral matter.

Isotropic Carbon - Amorphous, non-oriented insoluble material.

Anisotropic Carbon - Insoluble material which exhibits an ordered
                       or layered structure as evidenced by optical
                       microscopy under polarized light.

     Mesophase - Anisotropic carbon which exhibits the charac-
                   teristics of a liquid crystal. It can occur
                   in several forms such as the following:

                   ● *Small spheres or domains*

                   ● *Larger or coalesced spheres*

                   ● *Mosaics - large areas of coalesced*
                     spheres but segregated into irregular
                     shapes

                   ● *Flow-type coalescence - the entire*
                     mass has oriented into sheets or
                     needles

   Glassy Carbon - Small filaments of ordered carbon intertwined
                     to produce an overall appearance of disordered
                     structure - generally possesses high surface
                     areas.

---

particular, they discussed the sequential formation of flow-type
anisotropic carbon: soluble materials condense to first yield
small, spherical anisotropic domains, which coalesce to larger
domains, mosaics and finally fully-oriented massive flow-type
structures.

Large planar aromatic molecules form the initial anistropic
domains by interacting and aligning into an onion-skin-like
structure.

Non-planar molecules lead to the formation of isotropic carbon.
For the formation of flow-type or needle coke, two requirements
must be fulfilled:

1. Small soluble molecules must undergo a condensation reaction to form larger condensed ring molecules.
2. Rearrangement must be possible to produce a layered
structure.

"Both processes must take place simultaneously without mutual
interference to form anisotropic material. However, the condensation reaction brings about an increased viscosity of the
matrix with an increase in molecular weight, which should retard
free motion of molecules to form a layered structure. Thus the
condensation reactivity should be appropriate to satisfy two
contradictory requirements." (8-8(c))

If the condensation reactivity is too high, the reaction's
progress may stop at the isolated domain or mosaic stages due to
restricted free motion. Similarly, if the starting material is a
mixture of compounds (which usually is the case) and the condensation reactivities are different, each will form its own separate series of domains. When the mixture consists of both
planar and non-planar molecules with near equal reactivity,
isotropic or very small domain anistropic carbons will result due
to the disorder introduced by the non-planar molecules.

An added complication is that impurities in the feed may
adhere to the exterior surface of the isolated domains and prevent
further coalescence.

Model compounds have been used to identify which chemical
structures condense to form flow-type anisotropic carbon. Most of
this work, however, was conducted at higher temperatures (1000°F)
and pressures (14,000 psi) than are employed in coal liquefaction
(8-5,8-6,8-7) or was conducted in the presence of strong Lewis
acid catalysts (8-8). Nevertheless, the mechanisms involved are
believed relevant in SRC-type processes, also. The model compounds most widely studied have been simple condensed aromatic
rinç structures:

Benzene

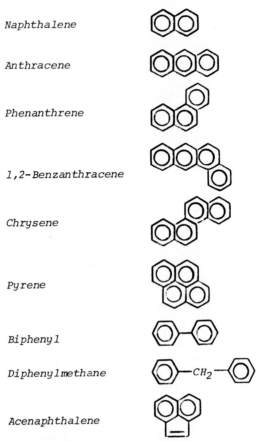

Naphthalene

Anthracene

Phenanthrene

1,2-Benzanthracene

Chrysene

Pyrene

Biphenyl

Diphenylmethane

Acenaphthalene

Thermally carbonizing such simple structures is very difficult; the necessary conditions to achieve the first step - anisotropic sphere formation - are shown in Table 8-2.

Other workers (8-6,8-11,8-12) have shown that the relative rates of pressurized carbonization of biphenyl/phenanthrene/ anthracene are 1/3/1200, respectively. This difference in rates has been attributed to the relative ease of condensation to form large planar sheets such as:

9,9'-Bianthryl   Meso-naphthodianthrene

Another view (8-7) is that the carbonization rates of simple aromatics are proportional to how easily the molecules can be excited from a singlet to triplet state and then polymerized to give structures such as:

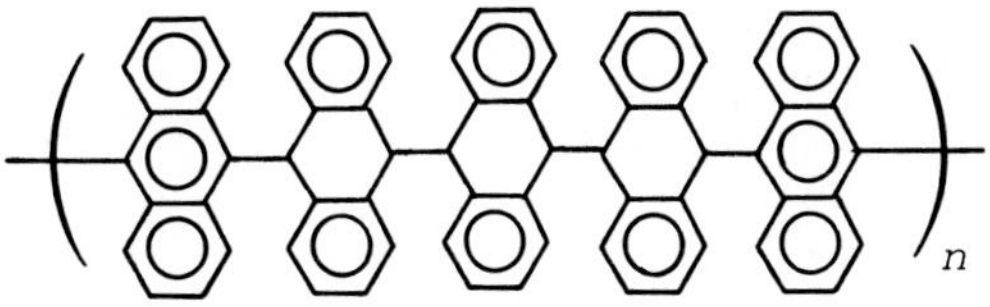

This point is further supported by this observation: the relative rates or carbonization of biphenyl/phenanthrene/anthracene are in line with the relative rates of addition of methyl radicals to biphenyl/phenanthrene/anthracene (5/27/820, respectively) (8-13).

High temperatures and pressures are not always needed for carbonizing, however. Many large condensed systems, such as the condensed anthrone-type dyes, form flow-type mesophase at atmospheric pressure and 970°F (8-4,8-14).

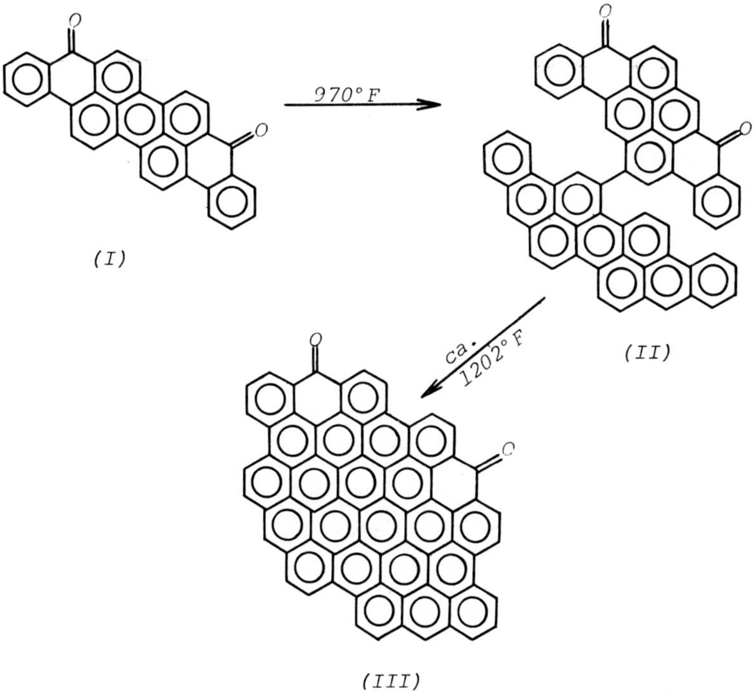

Chemical changes during carbonization of dibenzanthrone.

The temperature of mesophase formation can be further reduced by the addition of catalysts such as $AlCl_3$ (8-8(b), 8-8(d)) or alkali metals (8-8(c)). Compare the results in Table 8-3 for $AlCl_3$-catalyzed carbonizations at atmospheric pressure, with those of thermal carbonization at very high pressure given in Table 8-2 for naphthalene, biphenyl and anthracene. In each case, over 85% of the compound charred at much lower temperatures and pressures when $AlCl_3$ was present.

Carbonizations catalyzed by alkali metals require higher temperatures (930°F) than $AlCl_3$-catalyzed conversions. The sensitivity of the mesophase formation to catalyst concentration is shown in Figures 8-1 and 8-2.

The proposed mechanisms for the catalyzed condensations which yield mesophase are shown below (8-8(c)):

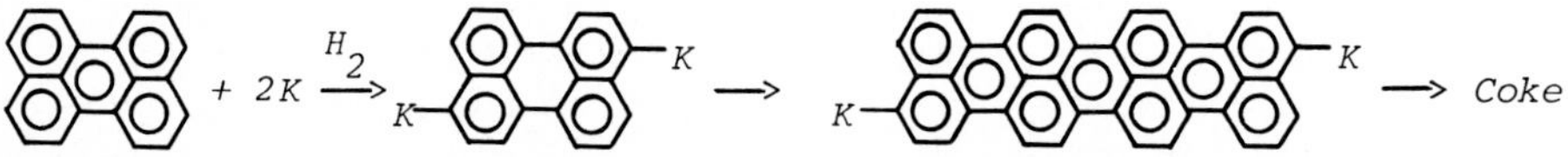

Alkali-Metal-Catalyzed Condensations

$$AlCl_3 + H_2O \rightleftharpoons H^+ AlCl_3(OH)$$

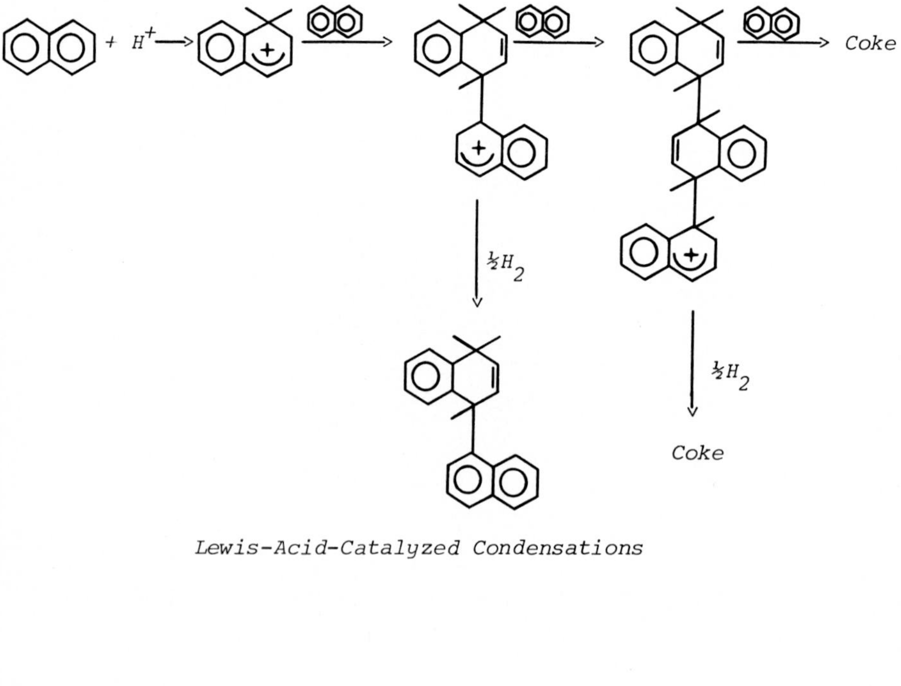

Lewis-Acid-Catalyzed Condensations

TABLE 8-2.   Thermal Treatment of Aromatic Hydrocarbons[a]
(Two Hours at Indicated Temperature)

	No Coke Formed	Anisotropic Spheres Formed
Naphthalene		
Temperature, °F	1067	1112
Pressure, psi	18,860	18,860
Acenaphthalene		
Temperature, °F	896	932
Pressure, psi	33,370	39,174
Biphenyl		
Temperature, °F	1022	1112
Pressure, psi	29,020	30,470
Phenanthrene		
Temperature, °F	980	1022
Pressure, psi	29,020	42,080[b]
Anthracene		
Temperature, °F	—	932
Pressure, psi	—	24,670

[a]Marsh, H., et al. (1973). Fuel 52, 235.

[b]Coalesced mosaic.

An interesting observation in the $AlCl_3$-catalyzed carbonization is that pyrene, which is somewhat less reactive than other
less condensed aromatics, does not form flow-type anisotropic
carbon (e.g., the precursor of needle coke) but stops growth at
the mosaic or large coalesced domain forms. A reason for this
behavior has been proposed (8-8 (b)): at the higher temperature
needed to initiate the reaction, secondary condensation occurs
too quickly, restricting free motion and molecular alignment.

TABLE 8-3. *Yield of Benzene-Insoluble Material (essentially mesophase) From Simple Aromatics in AlCl$_3$-Catalyzed Carbonizations[a]* (.1 mol % AlCl$_3$, holding time two hours, 1 atm.)

Samples	572°F	716°F	932°F	2282°F
Benzene	–	0	–	–
Naphthalene	89	98	100	100
Anthracene	67	88	–	–
Phenanthrene	66	69	–	–
1,2-Benzanthracene	–	71	–	–
Chrysene	–	82	96	100
Pyrene	–	30	95	100
Biphenyl	41	89	–	100
Diphenylmethane	87	–	–	–

[a]Mochida, I., et al. (1975). Carbon 13, 135.

A solution to this mismatch in rates is to partially hydrogenate the aromatic nucleus or substitute it with alkyl groups. This increases the reactivity such that condensation can occur at a lower temperature, and slows down secondary condensation reactions so that alignment can occur. Thus flow-type mesophase can be produced. Needle coke has been produced by partially hydrogenating SRC and coal tar pitches (8-8(c)-(e)).

Physical solubility of the solvent components is also important in producing desirable residues. The term "compatibility" has been used to describe a combination of physical solubility, relative reactivities, and the ability to align molecules in mixtures. With this approach, optimizing the production of flow-type mesophase and therefore needle coke has been achieved for mixtures of different pitches and their fractions (8-8(e)-(f)). By contrast, if a mixture consists of mutually miscible materials but one of them is not reactive to condensation, it acts as a diluent and isolates growing domains, thus preventing coalescence (8-5(b)).

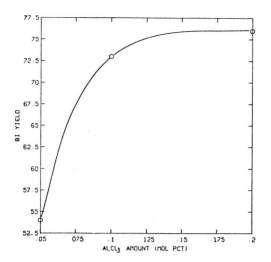

FIGURE 8-1.   Influence of the AlCl$_3$ Amount on the Carboniza-
tion Rate of Chrysene (2 hr holding at 716°F).   Benzene Insoluble
Yield (%) = (Benzene Insoluble Percent of Carbon) x (Carbonization
Yield)

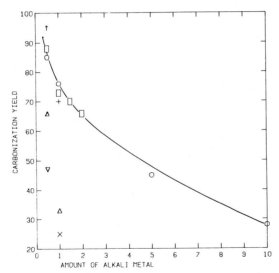

FIGURE 8-2.   Carbonization Yields of Aromatic Hydrocarbons
vs. Added Amount of Alkali Metal.   Heat Treatment 2 hr at 932°F.
∇,Anthracene + Li; Δ, Anthracene + Na; O, Anthracene + K; ↑,
Anthracene + K (in Diglyme);☐ , Pyrene + K; X, Naphthalene + K;
+, Naphthalene + K (in THF).

Inert solids present in carbonizing solutions can have one of two effects (8-5(b)). Small particles (colloidal size) severely restrict the mesophase growth process (8-5(b),8-10), but larger surfaces promote coalescence to form anisotropic units such as seen in metallurgical cokes (8-5(b)). The small inert particles are believed to adhere to the growing mesophase surface and prevent coalescence.   Such interactions are important in coal liquefaction where large quantities of finely dispersed unreacted coal or mineral matter are present.   A large portion of the observed mesophase in coal liquefaction residues is associated with mineral matter or coal macerals (8-3).

The above discussions centered on coke or char formation from aromatic hydrocarbons.   We will now briefly cover the coking behavior of heterocyclic or functionalized aromatics.   Some studies have recently been reported by H. Marsh (8-5(b)-(c)) on high pressure carbonization behavior of a variety of oxygen, nitrogen, and sulfur containing compounds.

As mentioned above, large fused-ring ketones (organic dyes) such as dibenzanthrone, pyranthrone, and flavanthrone form mesophase easily.   These large fused ring systems are often synthesized by condensation reactions of smaller fused ring systems as shown below for dibenzanthrone (8-15). These large molecules have difficulty in achieving high degrees of alignment and are often limited in the development of completely coalesced mesophase, but mixing small and large molecules has been shown to be synergistic for coalescence (8-15).

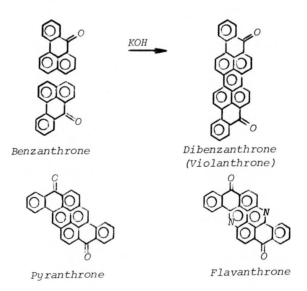

Benzanthrone   Dibenzanthrone
                                            (Violanthrone)

Pyranthrone   Flavanthrone

Coke usually contains heteroatoms because its common precursors are not pure aromatic hydrocarbons. Partially carbonized coke-oven-pitch has been reported to contain up to 1.8% oxygen (8-4). For coal tar pitches, 5-8% heteroatoms can be accommodated without disrupting the anisotropic structure; higher concentrations (which contain as much as 5% nitrogen or 20% sulfur) limit the domain size (8-5(a)). Coalesced anisotropic residues have been produced under high pressure carbonization from mixtures of acenaphthalene and heteroatom-containing compounds. Even higher heteroatom contents were achieved for the pure heteroatomic materials. The reactivity for heterocyclic aromatics relative to aromatic hydrocarbons toward carbonization is very sensitive to the particular type of compound.

Adding specific oxygen-containing compounds reportedly improved the production of needle coke (8-16). For example, fluorene and carbazole form poorly graphitizing carbons alone but in admixture with phthalimide, phthalic anhydride, or pyromellitic anhydride, graphitization was improved. The mechanism by which this improved behavior occurred is not clearly understood, but other workers (8-5(c)) have shown that extensive phenyl substitution of both fluorene and carbazole occurs with the added compounds.

Phenolic materials are prevalent in coal tar pitches and SRCs. Under high pressures, carbonization of multi-ring phenols has been shown to lead to mesophase formation (8-5(c)). However, at low pressure, phenols are reported to induce condensation reactions to form isotropic carbon at temperatures well below that at which mesophase would normally form (8-4). Similarly, oxidation of coking feeds leads to isotropic carbon formation. These results offer some explanation as to why the benzene-soluble portions of SRC produce good needle cokes whereas the benzene-insoluble portions do not (8-8(c)). The highly functional components of SRC (phenols) have been shown to be predominantly benzene-insoluble (8-1). In fact, if phenols and bases are extracted from pitch, spherical mesophase forms at much lower temperatures (8-4) than required for the whole mixture.

Mesophase formation has been a common experience in operating SRC process-development units (8-3) and temperatures must be carefully controlled in preheaters to prevent fouling. In the catalytic hydrogenation of coal without a solvent, the rate of mesophase formation is reported as inversely proportional to the rate of stirring (8-9(a)). Thus, to prevent mesophase formation, the reacting mass must have access to hydrogen. The hydrogen content of the coal liquid is also critical, as can be seen by comparing the yield of fixed carbon to the elemental composition of SRCs as shown below (8-8(c)).

| SRC H/C mole ratio | .79 | .76 | .73 | .68 |
| Yield of fixed carbon | 35.6 | 43.8 | 52.8 | 58.4 |

The authors also showed that benzene-insoluble portions of SRCs can be converted from poorly graphitizing materials to good precursors of needle coke through mild hydrogenation (8-8(a)). This was presumed to be due to lower rates of condensation reactions compared to the molecules' alignment to a layered structure as the coke was formed.

It is commonly believed that the formation of mesophase during coal liquefaction is irreversible. However, it has recently been reported that a mosaic-type mesophase (H/C = .4) can be hydrogenated catalytically ($SnCl_2$) to produce 14% C_1-C_5 gases and 33% benzene-soluble liquids. The conditions used were similar to those in coal liquefaction: 842°F, 1857 psi H_2, 90 minutes (9-9b).

This result indicates that a single mesophase particle is not necessarily purely polymeric but may be an assemblage of large molecules held together by physical rather than chemical bonds. Some older work also points in this direction (8-14(c)) for spherical mesophase could be completely dissolved in concentrated H_2SO_4 if carbonization temperatures of 1100°F were used.

The above information on the chemical properties and formation of isotropic and anisotropic carbons will be integrated with the following discussions of our work on the chemistry of coke formation during coal liquefaction.

The following general reactions of coke formation from SRC will be discussed, based on work at MRDC:

• Regressive reactions which produce a solid deposit in reactors. This occurrance is usually deleterious to the process and should be controlled to minimize the quantity of char.

• Decomposition of the coal liquids in delayed coking units, which can increase the amount of light products produced in the overall liquefaction process. These reactions are technologically beneficial and are used in the Exxon Donor Solvent Process. The main goal in this instance is to increase the selectivity of the overall process toward light liquids.

• The use of coal liquids to produce a certain quality of coke with anisotropic optical properties, for example, needle coke which has specific precursors. Identifying the precursors of different types of coke and the specific conditions in which the carbonization reactions should be performed is very important. In this case the yield should be optimized for highly ordered coke.

II. COKE FORMATION STUDIES

Three experimental methods were used in this study to determine the quantity of char formed from different coal liquids or fractions thereof.

- Thermogravimetric analysis (MRDC).
- Carbonizing experiments in pressurized gold-tubes (Penn State).
- Reactions conducted in stirred autoclaves under typical liquefaction conditions (MRDC).

The optical properties of a number of residues from these experiments were determined at The Pennsylvania State University. The fractionation methods employed were based on solubility classes (benzene-soluble, benzene-insoluble) as well as liquid chromatographic fractionation method (SESC) to separate specific chemical classes.

A. Thermogravimetric Analysis of Coal Liquids (TGA)

When an SRC is heated up to 800°F under a helium atmosphere, about 40% of the initial weight of the SRC chars. To investigate the classes of compounds responsible for charring, similar experiments have been performed using various fractions of coal liquids. The benzene-soluble fraction (asphaltenes) of the SRC gave around 20% residue, whereas the benzene-insoluble fraction (asphaltols) yielded over 70% residue (Figure 8-3).
The thermogravimetric analyses of the different SESC fractions of SRC give similar results; SESC fractions 3,4,5 give small amounts of char at these conditions while SESC fractions 7,8,9 give over 60-70% char (see Table 8-4).
As can be seen in Figures 8-3 and 8-4, the major changes in the amount of residue obtained with this TGA technique occur between 350 and 450°C (662-842°F); after 500-550°C (932-1022°F), the amount of char remains essentially constant up to 1472°F. Some volatile compounds evaporate before the charring temperature is reached, however; if they had remained in the sample, the char yield at higher temperatures would probably have been higher. This supposition was proven correct by charring the same compounds in the pressurized gold-tube experiments (see below). In very high-pressure carbonizations, essentially all of the feed is converted to coke (8-5).
Figure 8-3 shows the results of the thermogravimetric analysis of a solvent-refined coal and its fractions both in helium and air. The sensitivity toward oxidation of the solvent-refined coal and especially of its benzene-soluble fraction is clearly evident (Figure 8-3(a) and 8-3(b)). The quantity of residue up to the combustion temperature of 1112°F is much higher when the TGA is performed in the presence of air than of helium. For example, at 932°F, 60% of the initial weight was still present as residue in air, compared with 20% of the initial weight in helium. The chemical nature of the two residues is most probably different as

D. Duayne Whitehurst *et al.*

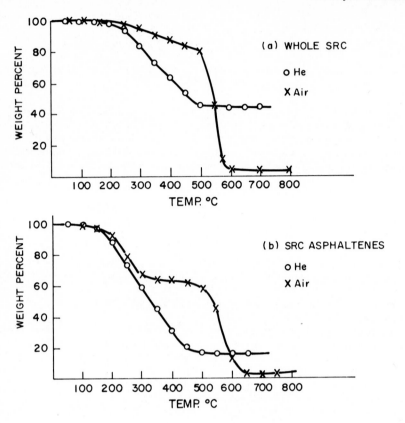

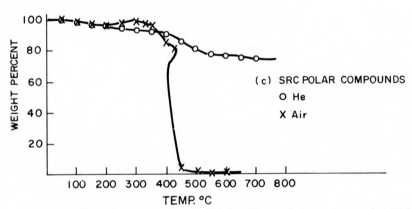

FIGURE 8-3. *Thermogravimetric Studies of West Kentucky SRC.*

TABLE 8-4.   Coke Yields in Thermogravimetric Experiments of
SESC Portions of Solvent-Refined Coals

Coal	Severity of the SRC Process[a]	SESC Fractions	Char Yield at 1472°F
Kentucky 9,14	LCT	3	30
		4	22
		6	48
		7	75
		8	75
		9	75
Illinois #6 Monterey	LCT	8	Mesophase[b]
	SCT	8	44
	SCT	9	66

[a]LCT - long contact time.
 SCT - short contact time.

[b]Gas formation and sample decomposition prevented accurate determination of the char yield.

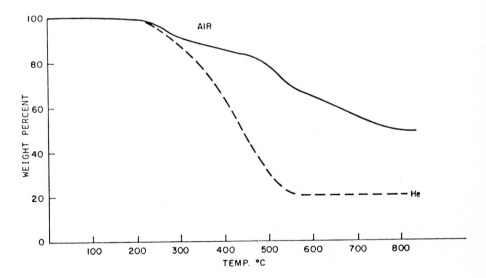

FIGURE 8-4.   TGA of Illinois SRC Benzene-Soluble Fraction.

TABLE 8-5.   *Atomic Ratio in the Volatile Matter*
Formed During TGA Experiments

	Monterey SCT SESC-8 Charring Temperature 1472°F	Monterey LCT SESC-8 Charring Temperature 932°F
Initial	$C_{100}H_{84}Het_{15}$ [a]	$C_{100}H_{80}Het_{14}$
TGA Residue [b]	$C_{32}H_{6}Het_{3}$	$C_{72}H_{38}Het_{9}$
Evolved	$C_{68}H_{78}Het_{12}$ $(C_{1}H_{1.15}Het_{0.18})$	$C_{28}H_{42}Het_{5}$ $(C_{1}H_{1.5}Het_{0.18})$

[a] Het = total heteroatoms (O+S+N) content.

[b] *Residue calculated for % char formed in TGA.*

the residue in air should contain oxidized compounds. The benzene-insoluble fraction of the SRC was less sensitive toward
oxidation, presumably due to its high phenolic content which
could act as an antioxidant.

A general observation based on the above data is that the
highly polar compounds (polyphenols of the SESC fractions 7,8,9)
and the oxidation products of the monofunctional compounds (SESC
fractions 3,4,5 or the benzene-soluble part of SRC) produce larger
amounts of char than the unoxidized benzene-soluble fraction of
SRC. This result is consistent with the reported yields of cokes
from the carbonization of oxidized and unoxidized coal tar pitches
(8-4).

Additional experimental data based on the TGA method are presented in Tables 8-5 and 8-6, and show that the amount of heteroatoms remaining in the char is substantial. The hydrogen content
decreases with an increase in the temperature of charring. Based
on the analyses presented in Table 8-5, the aromaticity of the
evolved species probably increases with increasing temperature.
The overall C/H ratio in the evolved species is 1/1.15 for charring to 1472°F.

Another important observation (Table 8-6) is that coke with
anisotropic optical properties is obtained at atmospheric pressure
in certain cases (for example, SESC fraction 8 of our Illinois #6
(Monterey) long time SRC). This fraction is benzene-insoluble,
thus one would not expect anisotropic carbon to be produced since
studies on the carbonization of the benzene-insoluble fraction of

TABLE 8-6. Elemental Analysis of Some SESC Fractions of SRC and
of Their Residue After Thermogravimetric Analysis

Sample	Elemental Analysis			% O+S+N	Optical Properties[e]
	%C	%H	%O		
Monterey SCT[a]					
Fraction SRC-SESC-8					
Initial	78.6	5.5	11.7[c]	15.9[c]	
Residue after TGA at 1472°F (49%)	87.7	1.4	-	10.9[d]	Isotropic
Monterey LCT[b]					
Fraction SRC-SESC-8					
Initial	79.8	5.3	11.7[c]	14.9[c]	
Residue at 932°F (~70%)	82.4	3.6	-	14.3[d]	Anisotropic

[a] Short contact time SRC (4 minutes, 800°F).

[b] Long contact time SRC (90 minutes, 800°F).

[c] Direct determination.

[d] By difference.

[e] Optical analysis of the TGA residue was performed at Penn State University by G. Mitchell.

coal tar pitches report only isotropic carbon formation (8-5(a)).
Some additional data on the TGA charring of selected model compounds will be presented later during the discussions of the
relationship of the chemical structure to the yield of char.

B. Carbonization Experiments in Pressurized Gold Tubes

Penn State has used gold-tube carbonization techniques to study the propensity of different substances or mixtures to form char (8-3). The procedure, as described in their report follows:

"The so-called 'gold-tube technique' involves heating small amounts of precursors, SRCs, fractions, etc. isothermally under an externally applied pressure. About 250 mg of sample is placed in gold tube (0.5 mm ID x 5 cm long) which is already sealed at one end. The sample and tube are evacuated and cooled in liquid N_2. After the introduction of a gaseous atmosphere, the tube is pinched, welded and subsequently tested for leaks. The sealed tube is placed in a hydrothermal pressure apparatus, pressure applied (5000-20,000 psi) and the autoclave rapidly heated to some desired temperature (425-500°C; 800-932°F). At the end of the run pressure is released and the heat-treated sample removed. Yield of semi-coke is then defined as the amount of pyridine insolubles formed during the treatment. Selected samples are examined by optical microscopy and Fourier Transform Infrared (FTIR) spectroscopy."

In a joint research program between MRDC and Penn State, the standard conditions used for the samples were one hour, 450°C (842°F) and 5000 psi of N_2. When mixtures of two fractions or of a fraction (or SRC) and an additive were used, the mixture was heated one hour at 350°C (662°F) prior to heating at 450°C (842°F) to ensure homogeneity.

The technique was applied to different SRC samples, their benzene-soluble and insoluble parts, as well as some SESC-chromatographic fractions (Table 8-7) which MRDC provided. The results of the gold-tube experiments using these samples are given in Tables 8-7 and 8-8. The role of additives, mainly tetralin (an H-donor) and aromatic hydrocarbons, on the charring of coal liquids was also studied in the gold-tube experiments (Table 8-9 and 8-10).

The data in Table 8-8 give the amount of char formed by different SRCs and their benzene-soluble and benzene-insoluble fractions. The quantity of char formed in these experiments and in the TGA experiments is practically the same for the benzene-insoluble fraction. Major differences were observed, however, for the benzene-soluble fraction and for the whole long-contact time SRCs. As the data in Figure 8-3 indicate, the quantity of char formed from the benzene-soluble fraction of long-contact time SRC was around 20% in TGA experiments while the value was only 0-2.5% in the gold-tube experiments.

TABLE 8-7. *Yield of Pyridine-Insoluble for Whole SRC's and Their Benzene-Soluble and Benzene-Insoluble Fractions Subjected to Standard Gold-Tube Conditions[a]*

Precursor	Pyridine Insolubles In Gold-Tube Products Wt %	Calculated Value of % Char of Whole SRC from % Char Fractions	Calculated-Experimental
Illinois #6 (Monterey) SC-SRC[b]			
Whole SRC	40	47.3	+7.3
Benzene Solubles (39% of whole)	2.5		
Benzene Insolubles (61% of whole)	76		
Illinois #6 (Monterey) LC-SRC[c]			
Whole SRC	0.3	16.8	+16.5
Benzene Solubles (78% of whole)	1.8		
Benzene Insolubles (22% of whole)	70		
Illinois #6 (Monterey) LC-SRC			
Whole SRC	0.1	13.6	+13.5
Benzene Solubles (79% of whole)	0		
Benzene Insolubles (21% of whole)	65		
Amax (Wyodak) SRC			
Whole SRC	3.0	22.3	+19.3
Benzene Solubles (71% of whole)	0.2		
Benzene Insolubles (29% of whole)	76.3		
Amax (Wyodak) SRC			
Whole SRC	23.9	33.5	+9.6
Benzene Solubles (61% of whole)	0.7		
Benzene Insolubles (39% of whole)	84.5		

[a] Penn State Univ. experimental data. [b] SC-SRC = short contact time SRC. [c] LC-SRC = long contact time SRC.

TABLE 8-8. *Elemental Analyses, Average Molecular Weights,
and Weight Percentages for Mobil SESC Fractions of
Illinois #6 (Monterey) SRC's Which Were Used
for Gold-Tube Experiments*

Frac-tion	C	H	O	N	S	Molecular Weight[a]	Wt %	Phenol Content[b] meq/g
Illinois #6 (Monterey) Short Contact Time SRC								
3	87.9	6.9	2.35	0.87	1.91	410	9.3	~0
4	83.0	6.6	6.8	1.1	2.4	490	18.5	1.4
5	79.3	6.7	10.4	1.4	2.0	680	13.6	1.4
7	80.5	5.8	8.8	2.1	2.3	1700	3.4	2.2
8	75.3	5.7	15.2	1.9	1.6	735	19.9	1.4
9	80.5	5.4	7.8	4.0	1.9		29.2	2.0
Total SRC								
	82.3	6.3	7.7	1.5	1.9		100	1.7
Illinois #6 (Monterey) Long Contact Time SRC								
3	89.2	6.0	2.7	1.3	0.8	480	27.9	~0
4	84.3	5.8	6.3	2.6	0.9	492	27.5	1.4
5	82.2	6.2	7.8	2.9	0.9	620	9.6	1.7
7	86.1	5.1	4.5	2.6	1.0	1050	5.3	1.0
8	83.8	5.4	7.0	2.5	0.8	1395	12.5	0.9
9	84.0	5.1	6.3	4.0	0.8		3.9	1.1
Total SRC								
	87.4	5.8	3.5	1.7	1.0		100	1.2

[a] *Vapor phase osmometry.*

[b] *These values may be low; the method used does not quantitatively detect hindered polyphenols. Acetylation indicates somewhat higher values.*

Table 8-9. Yield of Pyridine Insolubles for Various Mobil Fractions
of Illinois #6 (Monterey) SRC's After Gold-Tube Experiments
(Standard conditions: 1 hour, 842°F, 5000 psi N_2)[a]

SESC Fraction	Pyridine Insolubles After Gold Tube Treatment (Wt %)		
	LC-SRC[b]	SC-SRC[c]	
3	6.6	2.8	Polar aromatics, nonbasic N.O.S.
4	66.7	53.4	Monophenols
	64.5	51.9	
5	4.3	0.0	Basic N heterocyclics
7	73.5	66.5	Polyphenols
8	75.4	64.9	
9	nr	77.5	Increasing O content and increasing basicity of N
		72.0	

nr = not received.

[a] Reference 8-3.

[b] LC-SRC = Long Contact Time SRC Illinois #6 (Monterey).

[c] SC-SRC = Short Contact Time SRC Illinois #6 (Monterey).

This difference may be explained by the high pressure (5000
psi) used in the gold-tube experiments which favors internal H-
transfer and inhibits condensation reactions. The high pressure
may keep all of the components of the SRC together in the liquid
phase, including those which can donate H and those in which
condensation reactions can occur. This explanation is also consistent with the difference in the observed amounts of char formed
in the whole SRC and the calculated percent of char (using a
linear combination of the percent of char formed by each fraction
and its relative concentration). In each case (Table 8-7) the

calculated value was higher than the experimental observed value; this indicates strong interactions between the components during char formation in the gold-tube experiments. It has been reported (8-5(a)) that the purpose of high pressure in carbonization is to restrain the coalescence of the mesophase.

The gold-tube technique was applied to the SESC-fractions of two Wilsonville solvent-refined coals (a short-contact time SRC,

TABLE 8-10. *Yield of Pyridine Insolubles for Various Mixtures[a] of Illinois #6 (Monterey) SC-SRC and Tetralin[b]*

Precursor	Added Tetralin (Wt %)	Pyridine Insolubles After Gold-Tube Treatment (Wt %)
Whole SC-SRC[c]	50	0.0
	16.7	1.6,1.9
	8.3	3.8,3.9
	4.5	11.5
	None	43
Benzene Insoluble Fraction of SC-SRC[c]	14.3	45.8
	9.1	59.8
	8.3	64.3
	4.5	70.6
	None	76

[a]*All samples preheated at 662°F for 1 hour at 5000 psi prior to treatment under standard conditions.*

[b]*Reference 8-3.*

[c]*Illinois #6 (Monterey) SC-SRC.*

SC-SRC,and a long-contact time SRC, LC-SRC); both SRCs were produced from Illinois #6 (Monterey) coal.

The elemental analyses of these fractions, their relative
amounts in the SRC and their number-average molecular weights are
given in Table 8-8. The results of the gold-tube experiments with
these fractions are given in Table 8-9. The results obtained with
the SESC fractions 7,8, and 9 are not surprising since they are
similar to the benzene-insoluble fractions presented in Table 8-8.
Concerning SESC fractions 3,4, and 5, three observations are
significant:

• SESC Fraction-4, monophenols, produces a large quantity of
char (53.4% in short-contact time SRC and 66.7% in long-contact
time SRC), while SESC fraction-3 produces only insignificant
amounts of char.

• SESC Fraction-5, like SESC-4, contains phenols. In addition,
it contains basic nitrogen compounds and has a higher H/C ratio
and does not produce significant yields of char.

• The quantity of char for a mixture of SESC fractions 2,3,4
and 5 predicted from their relative proportions in SRC and separate char yields (assuming the char yield for fraction 2 to be
zero) are higher than actually observed for the whole benzene-
soluble fraction of the same SRC (Tables 8-6 and 8-8).

Precursor	% Char in The Benzene Soluble Part	% Char Calculated Based on Data of Fractions SESC 2-5	Calculated Minus Found
Ill.#6 (Monterey) SC-SRC	2.5	10.1	+ 7.6
Ill.#6 (Monterey) LC-SRC	0.1	20.6	+20.5

The above results indicate that the main species involved in
regressive reactions during SRC processing are the phenols
(SESC-4,7,8, and 9). In the benzene-soluble fraction of SRC, the
regressive reactions of SESC-4 are inhibited by the presence of
chemical species present in fractions 3 and 5. We believe that
the inhibition of the regressive reactions by fractions 3 and 5
may be due to their H-donor capacity. Structural studies show
that they contain hydroaromatic rings which could transfer hydrogen. In similar gold-tube studies, it was shown (8-3) that the
addition of tetralin to SRC also produces a dramatic decrease in
the quantity of char which is formed (see Table 8-10). Fractions
SESC 2,3, and 5 may help in reducing char formation in whole SRC
because of their good solubility properties for the rest of the
fractions.

Tetralin is not a good solvent for the highly polyfunctional compounds present in the SESC fractions 7, 8 and 9, which may explain the less dramatic reduction of char content if tetralin is added to the benzene-insoluble fraction of SRC (Table 8-10 and Figure 8-5). Some work in converting "asphaltols" in the presence of our standard synthetic solvent (8-1,8-17) is also consistent with this explanation.

The monophenolic fraction (SESC-4) of SC-SRC gives 53% char as compared to 67% char from the same fraction obtained from a LC-SRC (Table 8-9). Our data on the structure of different SRCs indicate a decrease in hydroaromatics with time in the SRC

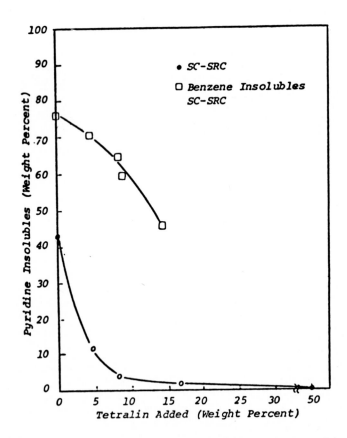

FIGURE 8-5. *Effect of Tetralin Addition on the Coking Propensities of Illinois #6 (Monterey) SC-SRC and Its Benzene-Insoluble Fraction (8-3)*

fractions. The result is that the monophenolic fraction in a long-contact time SRC will have less H-donor capacity to hinder its own condensation reactions. This is consistent with observations of other workers for the carbonization of coal tar pitches (8-8).

Data in Table 8-11 show that the quantity of pyridine-insoluble formed after the gold-tube treatment of SC-SRC is reduced not only by H-donor but also to a lesser extent by aromatic hydrocarbons such as biphenyl, anthracene, phenanthrene and naphthalene. Their action is also more effective on whole SRC than for the benzene-insoluble fraction alone. Several explanations are possible for this reduction in char. Since condensation reactions are at least second-order, dilution of the reactive species by an inert substance would have a significant effect on the rate of reaction. Also the benzene-insoluble fraction of the SRC is less soluble in naphthalene, resulting in incompatibility (8-8e) and the condensations occurring in a separate phase. Alternatively, such compounds could very well increase the interaction between H-donor benzene-soluble components (SESC fractions 2,3 and 5) and the char-forming fractions (SESC-4,7,8 and 9), for example, by H-shuttling.

Several recycle solvents were added to SC-SRC in the same conditions (Table 8-11). As those solvents contain both H-donors and large quantities of polyaromatic hydrocarbons, their effect of reducing the quantity of char is not surprising. The addition of an H-acceptor (benzophenone) caused a substantial increase in the percent of char formed.

C. Thermogravimetric and Gold-Tube Charring Experiments with Model Compounds

In a brief study on charring properties, different model compounds were also subjected to thermogravimetric analysis (Table 8-12). The majority of these compounds were either too volatile or as in the case of polystyrene or polyvinylnaphthalene decompose before charring. Only anthrarobin (a triphenol), 1,2-dimethylindole and polyfurfural alcohol charred in significant amounts.

J. K. Stille (Colorado State University) under subcontract with MRDC synthesized model compounds which have structural similarities with coal or coal products, but do not have alkyl substituents. These compounds were related to Stille's work on polymers of exceedingly high thermal and oxidative stability (8-18 through 8-22). The structure of the compounds, percent char formed in He at 800°C (1472°F) in a TGA, and percent char formed in gold-tube experiments are presented in Table 8-13. The quantity of char formed by these compounds appears to depend more on

TABLE 8-11. *Yield of Pyridine Insolubles for Mixtures[a] of Illinois #6 (Monterey) SC-SRC and Various Materials[b]*

Precursor	Solvent	Concentration of Additive (Wt%)	Pyridine Insolubles After Gold-Tube Treatment (Wt%)
Whole SC-SRC[c]	None	None	43
	Tetralin	16.7	1.6,1.9
	Biphenyl	14.3	11.1
	Anthracene	14.3	11.6
	Phenanthrene	14.3	14.1
	Naphthalene	14.3	19.1
	Benzophenone	14.3	50.6
	Recycle (Wyoming)	14.1	6.7
	Recycle (Ill. #6 (BS))	14.2	5.8
	Recycle (Ill. #6 (BS))	14.2	4.2
Benzene Insoluble Fraction of SC-SRC[c]	None	None	76.0
	Tetralin	14.5	45.8
	Naphthalene	14.3	73.8
	Benzophenone	14.3	88.9

[a]All samples preheated at 662°F for 1 hr at 5000 psi prior to treatment under standard conditions.
[b]Reference 8-3.   [c]Illinois #6 (Monterey) SC-SRC.

TABLE 8-12. *Thermogravimetric Analysis of Model Compounds*

Sample	Residue at 1472°F
Polystyrene	9.2
Poly-1-vinylnaphthalene	1.7
Chrysene	1.9
5,6,7,8-Tetrahydro-1-naphthol	4.0
Poly-tetrahydro-1-naphthol	8.0
6,13-Dihydrodibenzo-phenezine	4.0
Cholesterol	1.0
5,6-Benzoquinoline	2.8
4-Azafluorene	.5
Anthrone	.5
5,12-Dihydrotetracene	.0
Bibenzyl	1.8
5,6-Benzoquinoline	1.9
Anthrarobin	29.0
9,9'-Perhydrobiphenanthryl	3.0
1,2-Dimethylindole	19.5
9-Benzylidenefluorene	.8
Polyfurfural Alcohol	33.0

TABLE 8-13. Stille's Model Compounds Investigated for Char Formation

Compound	Mol. Wt.	% Char Yield TGA/He	% Char Yield Gold-Tube Experiment
Stille's Number 1	576	3	–
Stille's Number 2	484	35	0.7
Stille's Number 3	636	13.7	–

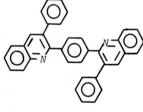

TABLE 8-13.(continued)

Compound	Mol. Wt.	% Char Yield TGA/He	% Char Yield Gold-Tube Experiment
Stille SON II-115	6508	65.0	--
Stille RMH-23	575	8.5	--
Stille WHB-225	628	10	0
Stille GLB-094	660	--	2.4

TABLE 8-13.(continued)

Compound	Mol. Wt.	% Char Yield TGA/He	% Char Yield Gold-Tube Experiment
Stille PS-153	Av. 2200	--	1.4

| Stille GLB-098 | 1248 | -- | 44.9 |

| Stille WHB-17 | $(420)_n$ | -- | 99.2 |

their molecular weight and their ability to form extensive condensed aromatic rings, when heated in an inert atmosphere, than
their phenolic OH content. For molecules which had similar backbone structures, a small reduction in coke yield was noted for
those which had no phenols (WHB-225). The internal H-transfer in
some compounds with hydroaromatic structures may have aided in
lowering the amount of char formed in the gold-tube experiments
relative to TGA measurements (WHB-225 and Stille #2). Lignin and
some model compounds related to it were also analyzed by thermogravimetric analysis. Only lignin itself gave any appreciable
amount of char (Table 8-14).

FTIR Spectroscopy of SRC fractions suggest that coking is
caused by chemical condensation of the alkyl and phenolic groups,
not by direct ring condensation to form polycyclic aromatics (8-3).
In addition, the type of alkyl functional group present is thought
to play a fundamental role.

The conclusions presented in The Penn State report (8-3) relating to the gold-tube carbonization experiments are summarized
below:

1. The pyridine-insoluble fractions of both SC-SRC and LC-
SRC of two bituminous coals include both an anisotropic
(mesophase semi-coke) and an isotropic phase when carbonized
under our standard condtions.

2. The amount of anisotropic semi-coke present in a pyridine-
extracted carbonized sample increases as the amount of
pyridine-insoluble material and soak time increases.

3. With soak time longer than two hours (up to five hours)
optical texture of the carbonized residue of SC-SRC changes
little.

4. Domains formed during carbonization of LC-SRC attain a
larger maximum size than those formed in SC-SRC.

5. Pyridine-insoluble residues from carbonization of SESC
fraction 4 are 100% mesophase semi-coke, both from LC and
SC-SRC. Fractions 7 and 8 from SC-SRC produce 100% mesophase-
semi-coke, and from LC-SRC perhaps somewhat less. The residues from these fractions of LC-SRC have much larger domains
than those of SC-SRC.

6. SESC fraction 9 does not form mesophase when carbonized
under our standard conditions; a granular texture develops.

7. Wyodak-Anderson coal, when heat-treated under pressure
with no hydrogen donor, becomes plastic and loses particle
integrity but does not become anisotropic. When a hydrogen
donor is added, anisotropic semi-coke is produced.

TABLE 8-14.   *TGA Char Yields of Lignin and Lignin Models*

Compound	Mol. Wt.	% Char Yield TGA/He
Sesamin[a]	382	1
Sesamolin[a]	398	1
Lignin		27

[a]These structures are similar to the backbone structures of lignins.

D. The Formation and TGA of Cokes From Fractions of SRC from
Autoclave Conversions Under Liquefaction Conditions

The above discussions concentrate on understanding the propensity of various components of coal liquids to form coke. The
experimental conditions differed from process liquefaction conditions, in that hydrogen gas and significant quantities of solvent were absent in the experiments. Large amounts of coal
minerals and fine particles of unreacted coal were also absent in
that work.

To further study the reaction sequence in the interconversion
of SRC fractions, high-pressure autoclave experiments were run
under typical liquefaction conditions. Selected coal fractions
were subjected to liquefaction conditions in the presence of a
synthetic recycle solvent. All the fractions produced coke
(pyridine-insoluble residues) at 800°F even in the presence of
solvents containing 40% tetralin, but the more highly functional
fractions produced more coke. The data from these conversions
are presented in Table 8-15 and are compared to the TGA coke
yields for these and similar fractions in Table 8-16.

A correlation is indicated between the TGA and SRC processing
coke yields, - with higher fraction functionality, both TGA char
formation and liquefaction residue formation tend to increase.

The resultant TGA cokes of Illinois #6 (Monterey) LCT and SCT
SESC-8 and Wyodak-Anderson LCT SESC-8 were submitted to Penn State
for optical microscopic investigation. They found that only the
Monterey LCT SESC-8 formed anisotropic coke (indication of mesophase formation) to any significant degree. It does indicate,
however, that anistropic coke can be formed in high yield even
in TGA apparatus.

We can presently speculate that ordered coke under these conditions results only from structures which can orient into planar
structures in agreement with the conclusions of previous workers
(8-4 through 8-16). The fractions which formed amorphous coke
contained three-dimensional skeletal structures, which would most
likely be found in Wyodak-Anderson SRC (8-27). Such structures
may also be possible in Monterey SCT-SRC; increased processing to
produce Monterey LCT-SRC results in increased aromaticity and
planarity of molecules. It has been observed that Wyodak-
Anderson coal will not form anisotropic coke in the absence of an
external source of hydrogen (as in SRC processes) (8-3). Some
related studies were conducted in which solvents, or SRC fractions,
were reacted with synthetic solvent or a model hydrogen acceptor
(benzophenone) in autoclaves at 820-840°F. The samples studied
and the resultant coke yields are given in Table 8-17. These
data clearly indicate that reactions which remove hydrogen from

TABLE 8-15. Conversions of Asphaltenes and Asphaltols at 800°F in H_2 and Synthetic Solvent

Parent Coal	Illinois No. 6	Wyodak	Wyodak	Wyodak
Feed Concentrate	Asphaltols	Asphaltenes	Asphaltols	Asphaltols
Time, minutes	84	111	20	20
Products (MAF)[a]				
Gases (C_6-)	4.9	13.7	25.4	7.6
Solvent range	21.1	20.9	2.4	7.6
SRC range	61.1	58.2	62.4	65.3
Residue	12.7	6.4	9.7	19.5
SRC[a] (oils)[b]	25.6	34.6	7.9	12.6
(asphaltenes)[b]	23.0	26.3	36.9	35.6
(asphaltols)[b]	51.4	39.2	55.2	53.03
Wt % H-consumption (solvent)	2.2	4.7	0.6	0.6
Wt % H-consumption (gas) (at 20 minutes)	1.4	1.9	0.6	0.6

DETAILED SESC ANALYSES[a]	F[c]	P[c]	F	P	F	P	F	P
Fraction 1	0	5.5	0.2	1.4	0	1.8	0	0
Fraction 2	0	9.9	2.1	27.7	0	3.4	0	8.3
Fraction 3	0	20.5	2.4	10.9	1.2	5.5	4.7	8.6
Fraction 4	0.6	9.4	54.7	16.3	13.7	28.7	20.6	25.8
Fraction 5	1.5	6.7	30.3	9.1	16.1	10.9	12.2	10.9
Fraction 6	3.1	4.6	6.5	22.8	7.5	6.5	6.1	3.6
Fraction 7	6.4	10.3	2.0	5.1	27.3	7.1	17.5	10.2
Fraction 8	15.4	18.4	0.6	1.7	23.3	8.7	16.1	10.4
Fraction 9	73.0	14.8	1.4	5.0	17.7	27.4	20.6	23.4

a. Normalized to 100%. b. These data represent only the eluted materials. c. F = feed; P = product.

TABLE 8-16. Coke Yields in Liquefaction and TGA Experiments

Coal	Source* Severity	SESC** Fraction	% Aromatic H	% Aromatic C	Inert 1472°F TGA Char Yield Wt%	800°F Autoclave Run Time min.	800°F Autoclave Run Coke Yield Wt%
Kentucky	LCT	2	82	--	--	--	--
"	"	3	81	83	30	60	~5
"	"	4	63	74	22		
"	"	5	49	65			
"	"	6			48		
"	"	7					
"	"	8	60	73	~75	191	~16
"	"	9					
Wyodak	"	2	56	80	--	111	<8
"	"	3	48	64	--	111	6
"	"	4	47	--	--	111	10
"	"	5	47	53	--	110	7
"	"	6	36	--	--		
"	"	7	45	60	54	20	10
"	"	9	--	--	--	20	20

*Severity of pilot plant production of SRC that was fractionated. LCT = long contact time.
**These fractions were obtained via non-optimal preparative SESC and are not pure fractions; they are concentrates of the indicated fraction containing contaminants of the adjoining fractions.

TABLE 8-17.   Coke Formation from Solvents

Reactant	Added Component	Atmosphere	Temp. °F	Time Min.	Wt% Coke Yield	TGA Char 1472°F
Synthetic Solvent	None	H_2	840	90	0	
Synthetic Solvent	Benzophenone	Ar	840	90	0	
Synthetic Solvent	Benzophenone + DHP*	Ar	840	90	0	
6663	Benzophenone	Ar	840	156	21	
6663	Benzophenone + Tetralin	Ar	840	156	12	
6540	Benzophenone + Tetralin	Ar	820	90	3	
17447	Benzophenone + Tetralin	Ar	820	90	1	
Solvent-1**	Benzophenone + Tetralin	Ar	820	90	16	9
Solvent-2	Benzophenone + Tetralin	Ar	820	90	>70	30
Solvent-3 450-800°F	Benzophenone + Tetralin	Ar	820	90	34	
Phenols***	None	H_2	800	90	0	
" "	Benzophenone	H_2	800	90	25	
" "	Benzophenone	Ar	800	90	51	

*DHP = 9,10-dihydrophenanthrene.
**Fractionated heavy solvent from Burning Star coal (see Reference 8-2).
***Isolated from Kentucky 9, 14 solvent (6663).

even simple compounds strongly promote coke formation. Note in particular, that solvent 6663 in the presence of benzophenone yields 21% pyridine-insolubles at 840°F in 156 minutes. Under the same conditions, when 3% tetralin is included as a hydrogen source the coke formation is almost halved.

The sensitivity of coke formation to hydrogen gas is shown in the series of reactions of 450-800°F phenolic solvent components (See Table 8-17). These experiments show that a normally stable solvent component can be induced to form coke by H-abstraction during the coal dissolution. The extent of this coking is controlled by the availability of other H-donors in the solvent and the presence of H_2 gas. In situations of limited H availability, these metastable solvent components will provide H to satisfy the demands of the reacting coal and on doing so will form coke. This phenomenon is often observed as high balances (based on coal fed) and low apparent coal conversions (as conversion is defined as 100 x (1 - wt residue/wt coal feed).

E. Optical Examination of Gold-Tube Pyridine-Insoluble Residues of SESC Fractions and Benzene-Extracted Fractions of SRC

A general description of the gold-tube residues (cokes) from Penn State University (8-3) and the conditions under which they were formed is provided in Table 8-18. The various types of coke are shown in Figures 8-6(a),(b),(c) and 8-7(a), (b), (c).

SESC fractions 4A, 7, 8 and 9B, representing 71% of the Illinois #6 SC-SRC, produced enough pyridine-insoluble material to permit optical characterization. The pyridine-insoluble gold-tube residues of SESC 4A, 7 and 8 treated at the standard conditions (8-3) were composed entirely of anisotropic carbon. However, as seen in Figure 8-7(a) and Figure 8-7(b), the optical appearance of the residue from SESC-4A is considerably different from that of the residue from SESC-7. SESC-4A tends to produce large coalesced mesophase similar to that of the samples of whole Illinois #6 SC-SRC with a soak time of two to five hours, but with a larger domain size (5 to 25 μ for the fraction, compared with 5 to 10 μ for the whole). Residues from SESC-7 and SESC-8 have basically the same optical texture but have a smaller mosaic size (0.5-1.0 μ) than that from SESC-4A. The residue of SESC-9B is isotropic, with a trace of anisotropic material (0.1-0.5 μ). Apparently SESC 7, 8 and 9 have less ability to form distinct mesophase units under the test conditions (e.g., 5000 psi, 842°F, 1 hour). This helps explain the fact that the whole Illinois #6 SC-SRC, under these same conditions, produced only 18% (by volume)

TABLE 8-18. Microscopic Study of the Products From Gold-Tube Carbonization of SRC Fractions

Sample Identification	Source	Fraction	Wt % Pyridine Insoluble	Vol % Anisotropic Carbon	Size Range of Mesophase, μm
Illinois #6 (Mont.) SC-SRC	Mobil SESC Fractions	4A	53.4	100	Domains 5-25
		7	66.5	100	Mosaic 0.5-1.0
		8	64.9	100	Mosaic 0.5-1.0
		9B	72.0	0	Isotropic
Pyridine Insol- ubles	PSU Benzene Extraction	Benzene- Insoluble	NE	100	Mosaic 0.3-1.0
		Benzene- Soluble	NE	0	Isotropic
		Benzene- Soluble[a]	14.4	28	Mosaic 0.1-1.0, Domains 2-20
Illinois #6 (Mont.) LC-SRC Pyridine- Insolubles	Mobil SESC Fractions	4A	66.7	100	Domains 5-30, Flow Anisotropy 20-50
		7	73.5	93[b]	Coalesced Domains 5-30 Flow Anisotropy 5-10 w->50ℓ
		8	75.4	99[b]	Coalesced Domains 10-35 Flow Anisotropy 3-20 w->50ℓ

[a] Carbonized under 5000 psi at 842°F for 2 hours. All other runs carbonized under our standard con- ditions (5000 psi, 842°F 1 hour).

[b] Preliminary count, not duplicated.

NE = not extracted.   w = width.   ℓ = length.

anisotropic carbon in the pyridine-insoluble residue (8-3).
As previously stated, increased soak time for the whole SC-SRC
sample tends to increase the amount of anisotropic carbon in
the pyridine-insoluble residue as the domain size becomes
relatively stable. Presumably, with longer soak times the
whole SC-SRC can incorporate those fractions with less ability
to form large domains (7,8 and 9) in a planar aromatic frame-
work (as suggested by Marsh, et al., 8-5) similar to that of
the fraction 4A residue. This incorporation may produce the
large coalesced domains seen in the samples of whole SC-SRC
treated for two to five hours.

Optical characterization of heat-treated benzene-insoluble
and soluble fractions of Illinois #6 SC-SRC are reported in
Table 8-18. The benzene-insoluble fraction, when carbonized
under our standard conditions, produced a residue that was
similar to that described for SESC 7 and 8 of SC-SRC, i.e.,
small mosaic size (Figure 8-7(b)). Because the benzene-
soluble fraction produced an isotropic pyridine-insoluble
residue under the standard gold-tube conditions a longer soak
time (2 hours) was used to determine the ability of this
SC-SRC fraction to form mesophase units under more severe
conditions. As seen from Table 8-18 and Figure 8-7(c), the
2 hours soak time produced relatively large mesophase units
which were partially coalesced and dispersed throughout a
matrix of isotropic material. The sizes of domains produced
from this benzene-soluble SC-SRC component were similar to
those discussed for SESC-4A. The size and distribution of
anisotropic domains in the benzene-insoluble residue suggests
that this material is similar to SESC fractions 7, 8 and
probably 9, which are all highly phenolic.

Gold-tube pyridine-insoluble residues from SESC fractions
of the Illinois #6 LC-SRC are strikingly different from the
SC-SRC fractions from the same coal. Pyridine-insoluble
material was formed during carbonization of fractions 4A, 7
and 8 in sufficient quantities for optical characterization.
Although two of these fractions (SESC-7 and 9) were not
totally converted to anisotropic carbon, the domain size of
all the LC-SRC fractions is noticeably larger than that of
the SC-SRC residues. This observation lends weight to our
observations (discussed in Chapter 3) that the fractions of
LC-SRC are more aromatic than equivalent ones in the SC-SRC.
Figure 8-6(c) shows the large coalesced domains and flow
anisotropy which are typical of the LC-SRC fractions.

The observations suggest that interactions of the functional groups contained in these fractions restrict the
formation of mesophase units when they are combined in the

(a)   Partially coalesced and localized mesophase semi-coke in a matrix of isotropic Illinois No. 6 (Monterey) SC-SRC.

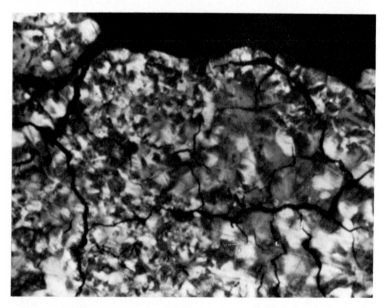

(b)   Coalesced domain (4-8 µm) and mosaic structure from the three hour gold-tube pyridine-insoluble residue.

(c) *Flow-band anisotropy and isotropic heat-treated SRC
from a five hour pyridine-insoluble residue. The isotropic phase
is seen in the lower left-hand quadrant of this photomicrograph.*

*FIGURE 8-6. Photomicrographs of Residues from Illinois #6
(Monterey) SRC Gold-Tube-Reactions. Reaction conducted at 5,000
psi, 790°F. Photomicrographed in reflected light, oil immersion,
using oblique-crossed Nicols, 1030X.*

unfractionated LC-SRC. The whole LC-SRC when carbonized under
the standard conditions produced very little pyridine-
insoluble material (0.3%, (8-3)). In contrast, the SESC
fraction produced substantial amounts of pyridine-insoluble
material and anisotropic carbon after carbonization under
standard conditions. Furthermore, individual LC-SRC fractions
(SESC 4A, 7 and 8) under less severe conditions, produced
anisotropic domains which were similar to those of the whole
LC-SRC sample treated for 5 hours. The amount of anistropic
material and size of anisotropic domains seen in the gold-
tube residues of the LC-SRC fractions were similar to those
seen in the gold-tube carbonization of the whole LC-SRC at
long soak times (8-3). However, at long soak times the whole
LC-SRC forms the same type of anisotropic carbon as seen in
the SESC fraction residues. Improved alignment of mesophase
during the carbonization of LC-SRC with time, implies that
molecular mobility is maintained and that little crosslinking occurs at this temperature (842°F) and in less than
5 hours.

This observation is consistent with lower functionality and
higher degrees of ring condensation in long-contact time SRC.

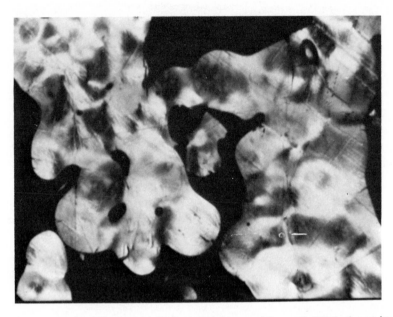

(a) Gold-tube pyridine insoluble residue of SESC-fraction 4A. Partially coalesced mesophase showing 6 to 20 µm domains.

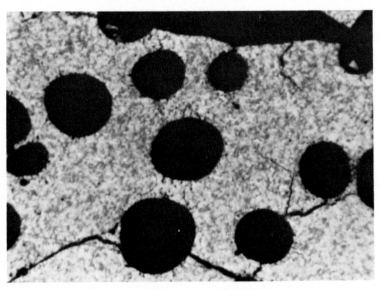

(b) Gold-tube pyridine-insoluble residue of SESC-fraction 7. Particle is composed of anisotropic mosaic ranging in size from 0.5-1.0 µm. Central vacuole is 18 µm across.

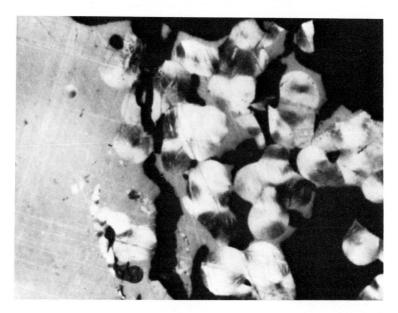

(c)   Gold-Tube pyridine insoluble residue of the benzene-soluble fraction. Photomicrograph shows spherical and coalesced mesophase (6-12 μm diameters) surrounded by an isotropic phase.

FIGURE 8-7.   Photomicrographs of Residues from Fractions of Illinois #6 (Monterey) SC-SRC Reactions. Reactions at 5,000 psi and 790°F. Photomicrographed in reflected light, oil immersion, using oblique-crossed Nicols, 970X.

F. Optical Characterization of Residues From Coal Liquefaction

Twenty-four samples were optically characterized by Penn State (8-3). The runs from which these pyridine-insoluble residues were derived may be divided into five general groups. Three of these groups were derived from runs with Wyodak-Anderson, West Kentucky 9,14 and Illinois #6 (Monterey) and (Burning Star) feed coals. The other two groups include runs using model compounds and solvents, and SRCs in various autoclave reactions. Brief descriptions of the runs (Tables 8-19 through 8-21) and of their residues are reported below.

1.   Reactions Using Wyodak-Anderson Feed Coal
All of these residues were derived from runs of relatively short reaction time performed by Princeton University (8-1). The residues were composed predominantly of unreacted or slightly

TABLE 8-19. Reaction Conditions for Wyodak Runs

Size and Pre-treatment of Coal, μm	Solvent	Solvent/ Coal Ratio	Temp. °F	Reaction Time, min.	Hydrogen Pressure, psig	Conversion, wt%	Autoclave Type
0-45 Demineralized	Mobil Synthetic	7.67	800	137.5	1300	80.98	Mobil, 300cc injection
45-75	Mobil Synthetic	13.4	878	1.25	1320	33.7	Princeton, 300cc injection
0-45	Mobil Synthetic	10.4	850	0.7	1510	45.4	Princeton, 300cc injection
45-75	Mobil Synthetic	14.3	865	0.5	1575	41.6	Princeton, 300cc injection
0-45	Mobil Synthetic	7.8	850	0.25	1500	53.5	Princeton, 300cc injection
45-75	Mobil Synthetic	10.96	850	0.15	1480	44.0	Princeton, 300cc injection

TABLE 8-20. Tube and Autoclave Reaction Conditions for West Kentucky 9,14 Runs

Sample No.	Pre-treatment of Coal	Solvent	Solvent/ Coal Ratio	Temp. °F	Time min.	Hydrogen[a] Pressure psig	Autoclave Type	Conversion wt%
T-1	NG	S-2 Wilsonville Solvent Fract.	2.99	752	47	1350	Tube	-68.2[b]
T-2	NG	S-3 Wilsonville Solvent Fract.	2.97	788	60	1100	Tube	57.7
T-3	NG	S-1 Wilsonville Solvent Fract.	2.92	788	60	1080	Tube	68.8
RN-1	0-600	Pyridine	5.00	795	60	500 (N_2)	Injection 300cc Autoclave	29.8

NG = Not Given.

[a] Tubes were filled with 300 psi hydrogen initially.

[b] Negative conversion indicates formation of additional solids.

TABLE 8-21. *Reaction Conditions for Runs with Illinois #6 (Monterey and Burning Star) Coals*

Sample No.	Size of Coal, μm	Solvent	Solvent/Coal Ratio	Temp. °F	Gas and Pressure psig	Time min.	Conversion wt %
RN-2	Monterey 75-425	SN6663 Wilsonville Solvent	4.89	804	Hydrogen, 594	90	84.24
RN-3	Monterey	Mobil Synthetic Substituting benzhydrol for tetralin	7.60	800	Hydrogen, ~900	180	Very low, charring noted
RN-4	Burning Star 75-212	p-Cresol	6.87	795	Nitrogen, ~500	2.8	53.3

altered coal particles (Figure 8-8(a)). Reflectance of some of these particles was increased relative to the feed coal. Large agglomerates also are common in these residues (Figure 8-8(b)). The following discussion appears in the Penn State report (8-3).

Vitroplast is rare in these short reaction samples, and none has formed the isotropic material resembling semi-coke which was present in Wyodak-Anderson filtration residue (8-3). These two substances are attributed (8-3) to the reformation of solids during liquefaction. Their absence suggests that regressive reactions may not occur during these short duration runs. Most organic particles in the residues did not appear to have undergone enough chemical alteration to have provided materials susceptible to regressive reactions. Physical breakdown of particles appears to lead only to increased agglomeration.

In contrast, the residue from a longer run (entry 1, 137.5 minutes, Table 8-19) with demineralized Wyodak-Anderson coal included a notable amount of anisotropic semi-coke. Mesophase semi-coke predominated in this residue, and recognizable coal components were rare. The mesophase coalesced to produce a domain size varying from 2 to 10 μ, and had incorporated angular fragments of isotropic material, presumably fusinite, semifusinite and/or vitroplast. Mineral matter was rare, but some pyrrhotite also is included in the coalesced mesophase.

2. Reactions Using West Kentucky 9,14 Feed Coal.

The conditions under which four West Kentucky 9,14 reactions were conducted are listed in Table 8-20. Three of these reactions were performed in small tubes with different heavy Wilsonville-solvent fractions (S-1, S-2, and S-3) under an initial hydrogen pressure of 300 psi. The fourth residue was obtained from an autoclave run with injection using pyridine as the solvent and a nitrogen atmosphere. Despite the difference in solvents and reaction conditions the residues from these runs were amazingly similar in appearance.

As seen in Table 8-20, there was actually a net gain in solids during test T-1, a 752° F tube reaction with S-2 solvent. Optical characteristics of this residue indicated that semi-coke formation definitely occurred during this run. Three particle types characterized this sample: mesophase semi-coke (Figure 8-9(a)), vitroplast, and wall-scale material (Figure 8-9(b)). Coalesced mesophase commonly has large domains up to a maximum of about 25 μ although the minimum size is near 1.0 μ. The vitroplast, as in the T-1 sample, usually is relatively highly reflecting. This com-

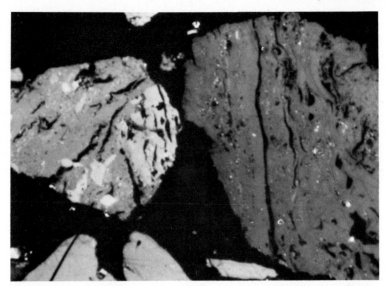

*(a) Contrasting particles of ulminite observed in the resi-
due from the reaction of a Wyodak-Anderson feed coal. Both particles are partially reacted, but the left-hand particle exhibits
higher reflections.*

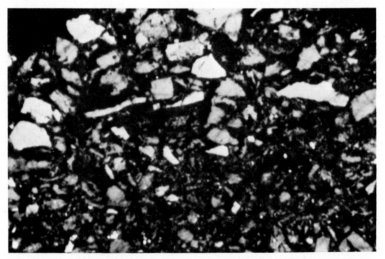

*(b) An agglomerate composed of fragments (0.5 to 1.0 μm and
4 to 8 μm) of slightly altered coal particles, observed in another
residue from Wyodak-Anderson coal processing.*

*FIGURE 8-8. Photomicrographs of Wyodak-Anderson Coal Conver-
sion Residues. Oil immersion and plane-polarized incident light,
970X.*

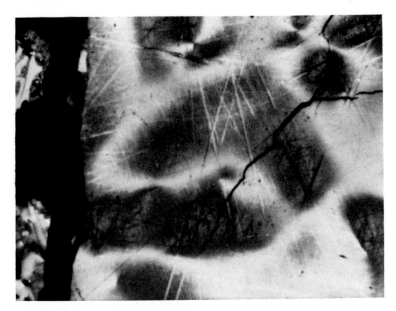

*(a) Large coalesced mesophase observed in the T-1 residue.
The large central domain is 66 μm along the long axis.*

*(b) An elongated, isotropic vitroplast-like component with
dispersed angular fragments of coalesced and spherical mesophase.
Particle shape suggests a wall-scale deposit in the T-1 residue.
The particle is 49 μm wide (short axis).*

(c) *Vitroplast with included aggregates of pyrrhotite, a
common component of the T-2 residue. This homogeneous and isotropic material exhibits several melt or flow structures. The
central aggregates of pyrrhotite are 18 μm across.*

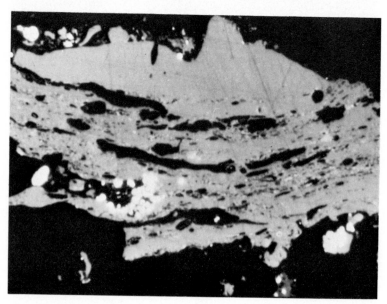

(d) *A highly degraded coal particle of higher reflectance
than was present in the original feed coal, observed in the RN-1
residue. The elongated central pore (dark slit) is 34 μm long.*

FIGURE 8-9. *Photomicrographs of West Kentucky 9,14 Coal
Conversion Residues. Oil immersion and plane-polarized incident
light, 970X.*

ponent can comprise a homogeneous matrix for aggregates of
coalesced mesophase. Particles of wall-scale, like that in
Figure 8-9(b) can be several tens of microns thick and
several millimeters long. Commonly they are agglomerates
of components such as minerals, fusinite, and semi-coke. In
the T-1 residue, the particles were similar to the stationary
deposits from Wilsonville described in The Pennsylvania State
University report (8-3) but without obvious alignment of components.

Samples from T-2 and T-3 (Table 8-20) were similar to the
T-1 sample although there was much less residual material
(i.e., loss instead of gain of solids from initial condition).
Vitroplast and isotropic wall-scale were the major components,
with mesophase semi-coke, minerals and inertinite forming the
remainder of the residue. Figure 8-9(c) shows a large
vitroplast particle with inclusions of pyrrhotite from the
T-2 residue. This vitroplast had a slightly lower reflectance
than that of vitroplast in the T-1 residue, but still dis-
played the typical characteristics of an isotropic, homogeneous material formed from West Kentucky 9,14 coal.

The residue from RN-1 was derived from reaction of the
West Kentucky 9,14 coal with pyridine in absence of hydrogen
gas. Petrographic studies of the residue revealed some
similarities with the tube reaction residues. The large
agglomerates of vitroplast which included particles and mesophase semi-coke were common. The vitroplast, as in the T-1
residue, was again relatively highly reflecting and appeared
as angular fragments throughout the residue. Mesophase semi-
coke was present as individual spheres and partially coalesced
masses of 3 to 8 µ diameter. Furthermore, there is some
evidence that coal particles were not reacted completely.
Figure 8-9(d) shows a remnant coal particle, partially reacted along the bedding planes and with reflectance enhanced
relative to that of the feed coal. Although this particle
type was not common in this residue, it was not seen in any
of the other tube-reaction residues using this coal.

The similarities among the West Kentucky 9,14 were
striking even though different solvents were used. Both
vitroplast and mesophase-derived semi-coke were common components. Again, it should be noted that the processing of
West Kentucky 9,14 coal results in the formation of abundant
vitroplast. This also was observed in the Wilsonville West
Kentucky 9,14 residues and fine particulates which plug filter
screens (8-3).

3. Reactions Using Illinois #6 (Monterey) and Illinois #6 (Burning Star) Coals

Two Illinois #6 feed coals were used in large autoclave runs with three different solvents as specified in Table 8-21. The Illinois #6 (Monterey) coal was used with a commercial recycle solvent from Wilsonville and a synthetic solvent in which benzhydrol (an alcohol) was substituted for tetralin. Also the Illinois #6 (Burning Star) feed coal was reacted with p-cresol in a nitrogen atmosphere. These residues cannot be optically compared because of the different solvents used in the reactions, but they will be described individually.

The Illinois #6 (Monterey) residue (RN-2) was largely composed of mesophase-derived semi-coke, unreacted or altered coal particles, and vitroplast. Figure 8-10(a) shows the large domains of coalesced mesophase; an attached layer of vitroplast or isotropic material displays the finely granular anistropic-mosaic structure of semi-coke. This particle is elongated in one direction and probably represents a wall-scale deposit. Although conversion to pyridine-soluble material was fairly high for this run (Table 8-21), unreacted or slightly altered coal particles are common. Even under these somewhat less severe reaction conditions and with relatively high conversion, regressive reactions (represented by mesophase semi-coke) and incomplete hydrogenation (unreacted coal particles) are observed in this residue.

The other run using Illinois #6 (Monterey) feed coal was reacted with a synthetic solvent in which benzhydrol was substituted for tetralin. During this reaction the anticipated hydrogen donor (benzhydrol) produced large quantities of a hydrogen acceptor which in turn removed hydrogen from both the coal and SRC, possibly promoting condensation reactions. Petrographic examination of the residue (RN-3) indicated that there was indeed little reaction. In Figure 8-10(b) a coal particle shows signs of partial reaction: semifusinite is seen in contact with a degraded and plasticized vitrinite of similar reflectance (middle of photograph) which was separated from a higher reflecting particle of fusinite. Partially reacted particles predominate in this sample. Some particle types which are suggestive of wall-scale formation also were present. However, they were poorly developed and relatively rare.

The third residue (RN-4), obtained from reaction of the Illinois #6 (Burning Star) with p-cresol, was composed of different materials than those of residues previously characterized. Figure 8-10(c) shows the vacuolated particles which are characteristic of this sample. These cenosphere

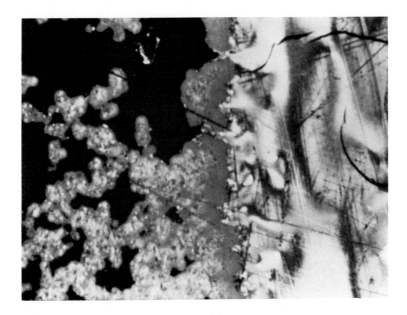

*(a)   A wall-scale particle observed in the RN-2 residue.
The particle is composed of large coalesced mesophase (right) with
a layer of isotropic material (center) that has developed finely
granular anisotropy (left, 0.5 to 1.0 μm).*

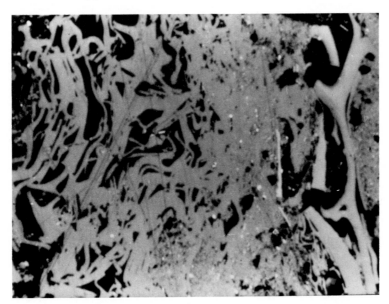

*(b)   An altered coal particle, which is a major particle-
type in the RN-3 residue. Semifusinite (left) has a reflectance
comparable to reacted and plasticized vitrinite (center) which
has separated from the higher reflecting fusinite (right).*

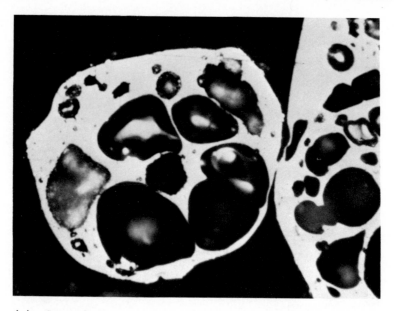

*(c)   Cenospheres are a major component of the RN-4 residue.
The large cenosphere (63 µm across) seen here has many gas
vacuoles (dark) surrounded by a thin layer of relatively high
reflecting and isotropic carbon.*

*FIGURE 8-10.   Photomicrographs of Residues from Illinois #6
Coals.   Oil immersion and oblique-crossed Nicols, incident light
970X.*

structures generally are composed of an isotropic, homogeneous substance with relatively high reflectance; darker
areas in the photomicrograph are gas vacuoles. Mineral
matter and fusinite also were present in this sample, but
are less common. This abundant formation of cenospheres was
not seen in any residue previously characterized. The fact
that little gas production was observed during this run
suggests that gases may have been formed and retained
internally in dissolving coal particles.

G. Reactions of SRC-SESC Fractions, SC-SRC, and a Coal Fraction

A few of the residues produced during the reactions aimed at
conversion of SRC-SESC fractions were also examined by optical
microscopy. Conditions of reaction and sources of feed materials
are reported in Table 8-22.

Residue RN-5 was derived from a run using Wyodak-Anderson
SRC-SESC fraction 9 with synthetic solvent containing 40% tetralin.
The following observations were reported by Penn State (8-3).

TABLE 8-22. Reaction Conditions for Autoclave Runs Using SC-SESC Fractions, SC-SRC and a Coal Fraction

Sample No.	Type of Feed	Solvent	Solvent/ Feed Ratio	Temp. °F	Hydrogen Pressure, psig	Time min.	Char wt %	Source of Feed
RN-5	Wyodak SESC-9	Mobil Synthetic	32.86	799	1277	20	19.5	Wyodak SRC-HRI SESC fraction prepared by Mobil
RN-6	Illinois #6 (Burning Star) SC-SRC	Mobil Synthetic + Pyridine	10.04	894	1422	4.9	17.0	Ill. #6 SC-SRC produced at Princeton in two short-contact runs at 427°F, 2000 psi, 2 minutes
RN-7	Illinois #6 (Burning Star) SRC-SESC-9	Mobil Synthetic	27.69	801	1285	191	16.1	Wilsonville Illinois #6 SRC
RN-8	West Kentucky 9,14 Pyridine Extract	Mobil Synthetic	5.98	801	1177	4.5	13.6	Pyridine extract produced at Mobil

    Because of the identity of the precursor, it is sur-
prising to see what appeared to be altered coal particles
and mineral matter in this residue. Much of the coal-derived
material was observed in agglomerates with inclusions of
pyrrhotite. The major particle type seen in this sample
was unusual. Figure 8-11(a) shows a fractured particle composed of an isotropic, gray reflecting and pitted material
with inclusions of mesophase semi-coke that appear to be
aligned within the isotropic phase. The mechanism of alignment of the mesophase semi-coke is not yet known, although
some of these particles appear to be elongated wall-sacle
deposits.
    Samples RN-6 and RN-7 were derived from processing an
Illinois #6 (Burning Star) SRC and SRC fraction-9 (Table 8-22).
Both of these residues were composed largely of anisotropic
semi-coke (mesophase and granular textured semi-coke). While
RN-6 contained only mesophase semi-coke in the form of angular
fragments and wall-scale particles, the RN-7 residue w s composed predominantly of granular textured semi-coke. Figure
8-11(b) shows an altered particle typical of RN-7, with a
finely granular anisotropic mosaic and many coal-derived
components (i.e., fusinite and mineral matter). Residue RN-6
has some partially reacted coal particles as well.
    Again, because the starting materials for these runs were
coal-derived products or fractions it is likely that the coal
particles found in these residues are contaminants.
    The final residue, RN-8 was derived from reaction of the
pyridine-soluble fractions of a West Kentucky 9,14 feed coal.
It was composed largely of an isotropic, gray reflecting
material which contained many vacuoles and some pyrrhotite
(similar to Figure 8-8(a)). Coal particles such as fusinite
and semi-fusinite also were present. Mesophase semi-coke was
rarely found, but when it was observed it exhibited a variable
domain size (0.1 to 30 μ) usually dispersed throughout the
isotropic phase.

H. Benzophenone and Solvent Reactions

    The series of reactions listed in Table 8-23 were pri-
marily aimed at measuring the hydrogen-donor capabilities
of commercial recycle solvents, recycle solvent fractions,
and synthetic solvent by using a standard hydrogen acceptor,
benzophenone. Another part of the study sought to determine
the amount of "char" formed from the depleted solvents during these reactions. The optical characteristics of these
"chars" or residues is valuable in establishing the role of
recycle-solvents in regressive reactions under typical

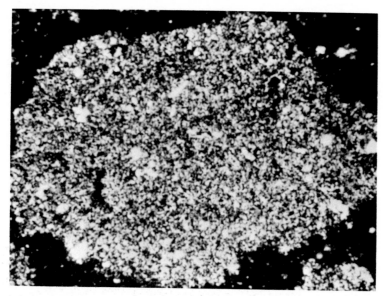

(a) An unusual alignment and elongation of mesophase domains
in an isotropic matrix seen in the RN-5 residue from the reaction
of a Wyodak SRC-SESC fraction No. 9 with synthetic solvent.

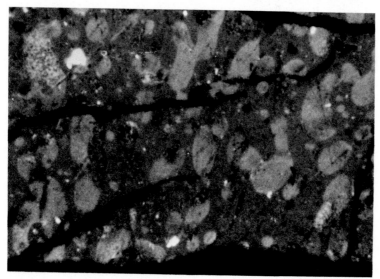

(b) Anisotropic semi-coke with inclusions of fusinite and
pyrrhotite, is commonly observed in this RN-7 residue from reaction of the Illinois #6 (Burning Star) SRC-SESC fraction No. 9
with synthetic solvent. The fine granular anisotropy (0.1 to
1.0 μm) is not apparent in the photomicrograph because of the
highly degraded nature of the particle.

FIGURE 8-11. Photomicrographs of Residues from SRC Fraction-
9 Reactions. Oil immersion, oblique-crossed Nicols, incident
light, 970X.

TABLE 8-23. Autoclave Reaction Conditions for Solvent-Benzophenone Runs

Sample No.	Solvent	Model Compounds	Temp. °F	Nitrogen Pressure, psig	Time min.	Amount of Char g/100g Solvent
RN-9	S-1 Wilsonville Solvent Fraction	Benzophenone	820	500	90	16.0
RN-10	SN17447 West Kentucky 9,14 Wilsonville Solvent	Benzophenone + Tetralin	820	500	90	0.7
RN-11	S-3 Wilsonville Solvent	Benzophenone + Tetralin	820	500	90	33.6
RN-12	SN6540 Wilsonville Solvent	Benzophenone	820	500	90	2.7
RN-13	Mobil Synthetic and 9,10-dihydrophenanthrene	Benzophenone	840	500	254	Very Little
RN-14	SN6663 Wilsonville Solvent	Benzophenone + Tetralin	820	500	157	11.8
RN-15	SN6663 Wilsonville Solvent	Benzophenone	840	500	250	21.0

(a) *Several types of anisotropic domains seen in coalesced mesophase in a wall-scale particle from the reaction of solvent S-3 with benzophenone (RN-11 residue). Oil immersion, oblique-crossed Nicols with incident light, 970X.*

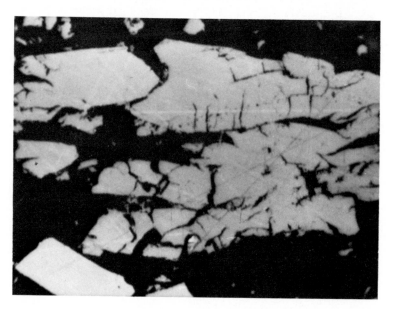

(b) *Fractured, homogeneous, and isotropic material seen commonly in the RN-14 residue from a reaction using SN 6663 solvent and benzophenone. Oil immersion, and plane-polarized incident light, 970X.*

FIGURE 8-12. Photomicrographs of Residues from Benzophenone Reactions.

liquefaction conditions.

Among the residue components observed in the samples were unreacted, partially reacted, and agglomerated coal constituents which included vitrinite, sporinite, fusinite and mineral matter.

These could either be the result of contamination from the autoclave or the solvent could have contained some unfiltered PDU coal particles. These particles were also present in samples RN-9, 10, and 13, although in much lower concentration.

Mesophase-derived semi-coke was observed in every residue except RN-13 and 14, and in two cases (RN-11 and 15), it was the major constituent. This semi-coke had a range of domain sizes and occurred as individual particles, angular fragments, and in wall-scale deposits (see Figure 8-12(a)). Mesophase semi-coke was seen in association with a low reflecting, isotropic material in most of these samples.

Another component found in significant concentration in several samples (RN-10, 11 and 14) was an isotropic, homogeneous, and highly fractured material seen in Figure 8-12(b). This particle type was optically similar to vitroplast, but because coal was not used in these reactions the term "vitroplast" would not strictly apply. The presence of such a material in this series of residues helps to substantiate the idea that not all of the vitroplast seen in commercial filtration and filter-feed residues is derived directly from coal particles.

The last residue component, a granular agglomerate composed of particles submicron in size to ∿5 μ, was observed only in the RN-13 residue. Examination shows the porous nature of these agglomerates. They contain identifiable particles of pyrrhotite, but the general chemical nature of the bulk of this material is unknown.

I.   Speculations on the Mechanism of Char Formation

The thermogravimetric studies, autoclave conversions and gold-tube experiments with coal liquids and their fractions discussed above indicate that phenolic OH is the main functionality related to char formation.

Based on FTIR spectra of residues obtained from gold-tube experiments, the Penn State researchers have suggested (8-3) that the mechanism of char formation for the phenolic fractions is the same as those proposed by Ouchi (8-24) for the pyrolysis of a phenol formaldehyde resin, as shown.

$$\text{Ph—OH} + \text{HO—Ph} \xrightarrow{-H_2O} \text{Ph—O—Ph} \tag{1}$$

$$\text{Ph—CH}_2\text{—Ph} \xrightarrow[+H_2O]{-H_2} \text{Ph—}\overset{\displaystyle O}{\overset{\|}{C}}\text{—Ph} \xrightarrow{-CO} \tag{}$$

(from rxn 1)

$$\text{Ph—Ph} \tag{2}$$

If reaction (1) is intramolecular, the result will be a less polar molecule; if intermolecular, the resultant molecule could also be less polar but will have a higher molecular weight. The intramolecular reaction cannot explain the increase in pyridine-insoluble fraction (as char is defined in gold-tube experiments). An appreciable amount of intermolecular water elimination and the formation of large molecules could be a cause of the decrease of solubility. However, simple water elimination cannot explain the high char yield of SESC fraction-4 (monophenols) in the gold-tube experiments or the decrease of the quantity of char due to H-donor or polyaromatic hydrocarbons addition. It was suggested (8-3) that in addition to water elimination, dealkylation and especially elimination of methylene bridges play an important role in char formation.

$$\text{Ar-CH}_2\text{-Ar} \longrightarrow \text{Ar-Ar} + Ch_x$$

It was found, however, that diphenylmethane type molecules are very stable (8-2). At the high pressure (5000 psi) involved in gold-tube experiments a reaction with gas formation may be even less favored.

More work is necessary to elucidate the complex mechanism of char formation in coal liquids. The elemental analysis of the starting material and of the pyridine-insoluble residue formed after the charring in the gold-tube experiments gave some indications of other possible (perhaps parallel) mechanisms.

The elemental analyses of a short-contact time SRC and its gold-tube pyridine-insoluble residues are shown below:

	%C	%H	%O
Calculated elemental analysis for fractions SESC 7,8 and 9 of an SRC*	78.5	5.5	10.7
SC-SRC	82.3	6.3	7.7
Pyridine-insoluble residue of SRC after gold-tube experiment	89.3	4.3	3.3
Calculated analysis of volatiles	50.9	4.1	6.4

*These are the major fractions responsible for char formation.

The above data indicate that elimination of small aromatic molecules and some oxygen-containing molecules must have occurred in the gold-tube experiment (empirical formula $CHO._1$).

To further understand the role of phenols in char formation, the phenolic -OH in an SRC was derivatized and the non-phenolic SRC reacted. One would expect lower char yields, but the opposite was observed because the derivative was thermally labile, creating a higher radical demand than the parent phenol, and thus, higher char yields. The details of this experiment and proposed mechanisms follow.

A short-contact-time SRC was acetylated with acetic anhydride in the presence of p-dimethylaminopyridine, a catalyst known to be effective for the esterification of hindered phenols (8-25). The acetylated SRC was free of OH groups as judged by its infrared spectrum. It was soluble in chloroform and also to a large extent in toluene. The results of the TGA and gold-tube experiments with this derivative and parent SRC are given in Table 8-24. The quantity of char both in TGA and gold-tube experiments increased as compared to the non-treated SRC. This result appears contrary to our conclusion that phenols increase "charring propensity". However, the derivative (an acetoxy group) is thermally unstable and easily fragments under liquefaction conditions via a Friedel radical reaction (8-26), producing phenoxy and acetyl radicals:

TABLE 8-24. TGA and Gold-Tube Experiments with a
Short-Contact Time SRC and Its Acetylated Derivatives

Sample	% Residue TGA, He, 1472°F	% Char Yield Gold-Tube Experiment
Whole SC-SRC	44	40
SC-SRC Fraction Soluble in Benzene	17	2.5
SC-SRC Fraction Insoluble in Benzene	62	76
Acetylated Whole SC-SRC	51	70

$$\text{(3)}$$

Both radicals are quite reactive. With the phenoxy radical
the most likely secondary reaction is H-abstraction; the acetyl
radical can proceed by the normal radical reactions or be
stabilized by reaction (4).

$$\text{(4)}$$

The phenoxy radicals formed in massive amounts will thermally
abstract H from practically any substrate, and they can consequently generate a large number of free radicals R· in the coal
liquids. These free radicals will then react further by the
least energetic pathway.

D. Duayne Whitehurst *et al.*

If sufficient H-donors are available then the H-transfer reaction is fast and the radicals are capped.

$$R \cdot \; + \; H\text{-donor} \longrightarrow R\text{-}H \tag{5}$$

If polyaromatic molecules acting as good physical solvents and H-shuttlers are present, the radical will stabilize by internal rearrangement or further aromatization.

$$R \cdot \longrightarrow R \; (\text{minus H or other small groups}) \tag{6}$$

If a large quantity of free radicals are formed, dimerization and polymerization can occur.

$$R \cdot \; + \; R' \cdot \longrightarrow R\text{-}R' \tag{7}$$

$$R\text{-}R' \; + \; \langle \bigcirc \rangle\text{-}O \cdot \longrightarrow (R\text{-}R')\cdot \; \text{etc.} \tag{7'}$$

$$R \cdot \; + \; \langle \bigcirc \rangle\text{-}O \cdot \longrightarrow R\text{-}O\text{-}\langle \bigcirc \rangle \tag{7''}$$

The ether formed in reaction (7") will be an end product if the bond is strong enough. Phenoxy radicals originating from phenols in the solvent will be incorporated into the product if dimerization occurs.

The polymerization reaction will be favored in the absence of sufficient H-donor and/or low concentrations of polyaromatic hydrocarbons which can aid in internal H-transfer reactions.

Water elimination from two phenolic groups can occur and this can explain some of the charring; however, we believe the reactions (7) and (7') are the major cause of charring.

This mechanism explains runs with heavy phenols isolated from recycle solvent. The heavy phenols did not char by themselves in small autoclave experiments (see the discussions above), presumably because they were predominantly monophenols which could stabilize by dimerization. They also contain hydroaromatic structures. However, when the heavy phenols were heated with coal under the same conditions, considerable charring occurred. This char may have been formed either from coal or the heavy phenols but it was observed that reaction with benzophenone in absence of coal also produced large amounts of char.

Coal liquids contain hydroaromatic structures, and during the polymerization reactions, more aromatic rings will undoubtedly be formed. Also, if aliphatic bonds in β positions from aromatic rings are present, or are formed during the heating, they will probably be broken due to their low bond strength (ca. 60 kcal/mole). These transformations were actually observed (8-23) when the residues formed after gold-tube experiments were examined by FTIR. Based on these data, it was concluded that charring resulted from condensation reactions of alkyl and phenolic groups with aromatic rings, rather than the fusing of rings to produce a more extensive polyaromatic ring system.

V. SUMMARY

Based on the above discussions, the structural features that increase carbonization or charring reactions are as follows:

- Increasing functionality (-OH, non-basic N)
- Increasing aromatic content
- Increasing molecular weight
- Decreasing volatility
- Catalysts
- Increasing H-scavengers
- Increasing rate of radical generation

Coke yields decrease with:

- Increasing H content
- Increasing H_2 pressure
- Increasing H-donors
- Dilution with inerts

REFERENCES

8-1.   D. D. Whitehurst, M. Farcasiu, and T. O. Mitchell,
         "The Nature and Origin of Asphaltenes in Processed
         Coals", EPRI Report AF-252, First Annual Report Under
         Project RP-410, February 1976.
8-2.   D. D. Whitehurst, M. Farcasiu, T. O. Mitchell, and J. J.
         Dickert, Jr., "The Nature and Origin of Asphaltenes in
         Processed Coals", EPRI Report AF-480, Second Annual
         Report Under Project RP-410-1, July 1977.
8-3.   P. L. Walker, Jr., W. Spackman, P. H. Given, A. Davis,
         R. G. Jenkins, and P. C. Painter, "Characterization of
         Mineral Matter in Coals and Coal Liquefaction Residues",
         EPRI Project 366-1, AF-832, December 1978.
8-4.   J. D. Brooks and G. H. Taylor, "Formation of Some
         Graphitizing Carbons", Chemistry and Physics of Carbon,
         Volume 4, (P. L. Walker, Editor), Marcel Dekker, Inc.,
         New York, 1968 p.234.
8-5.   (a) H. Marsh et al., FUEL 52, 205 (1973).
         (b) ibid., 235
         (c) ibid., 244
         (d) ibid., 254
8-6.   P. L. Walker, CARBON 10, 369 (1972).
8-7.   Pyrung Wha Whang, PhD Thesis "Pressure Effects on the
         Carbonization of Anthracene", The Pennsylvania State
         University, 1973.
8-8.   (a) I. Mochida et al., FUEL 53, 253 (1974).
         (b) I. Mochida et al., Bull, Chem.-Soc. Japan, 49,260
         (1976).
         (c) I. Mochida et al., CARBON 13, 489 (1975).
         (d) I. Mochida et al., CARBON 13, 135 (1975).
         (e) I. Mochida et al., FUEL 56, 49 (1977).
         (f) I. Mochida et al., Nenryo Kyokai-Shi, 56, 187 (1977).
8-9.   (a) M. Shibaoka and S. Ueda, FUEL 57, 667 (1978).
         (b) ibid., 673.
8-10.   Y. Sanada, T. Furuta, H. Kimura, and E. Hirok, Steiyu
         Gakkai Shih, 15, 936 (1972).
8-11.   M. D. Tilicheev, J. Appl. Chem (USSR), 12, 741 (1939).
8-12.   (a) C. R. Kenney et al., Ind. Eng. Chem. 46, 548 (1954).
         (b) C. R. Kenney et al., Proceedings of First and Second
         Carbon Conferences, University of Buffalo, p. 83 (1957).
         (c) C. R. Kenney et al., Ind. Eng. Chem., 49, (1957).
         (d) C. R. Kenney et al., Proceedings of Third Carbon
         Conference, Pergamon Press, p. 395 (1959).
8-13.   M. J. Szware, J. Chem. Phys., 23, 204 (1955).

8-14.   (a) J. D. Brooks et al., Contributors, Discussions,
        Residential Conf. Sci. Use Coal, 1958, pp. E39-E42.
        (b) J. D. Brooks et al., Carbon, 3, 185 (1965).
        (c) J. D. Brooks et al., Nature, 206, 697 (1965).
8-15.   L. F. Fieser and M. Fieser, "Topics in Organic Chemistry",
        Reinhold Publishing Corp., New York, p. 357, 1963.
8-16.   L. F. Isaacs, Carbon, 6, 765 (1968).
8-17.   M. Farcasiu, T. O. Mitchell and D. D. Whitehurst, Chem.
        Tech., 7, 680 (1977).
8-18.   J. F. Wolfe and J. K. Stille, Macromolecules, 9, 489
        (1976).
8-19.   S. O. Norris and J. K. Stille, Macromolecules, 9, 496
        (1976).
8-20.   W. Wrasidlo and J. K. Stille, Macromolecules, 9, 505
        (1976).
8-21.   W. Wrasidlo, S. O. Norris, J. F. Wolfe, T. Katto, and
        J. K. Stille, Macromolecules, 9, 512 (1976).
8-22.   J. K. Stille, R. G. Nelb II, and S. O. Norris,
        Macromolecules, 9, 516 (1976).
8-23.   P. C. Painter, Y. Yamada, R. G. Jenkins, M. M. Coleman,
        and P. L. Walker, Fuel, (1979) in press.
8-24.   K. Ouchi, Carbon, 4, 59 (1966).
8-25.   G. Hofle and W. Steglich, Synthesis., 619 (1972).
8-26.   S. Patei (editor), "The Chemistry of Carboxylic Acids and
        Esters", Interscience Publishers, New York, 1969, p. 352.
8-27.   G. D. Mitchell, private communication, January 1977.

Chapter 9

RELATIONSHIPS BETWEEN THE COMPOSITION OF RECYCLE SOLVENTS

AND COAL-LIQUEFACTION BEHAVIOR

I. INTRODUCTION

The chemical composition of a recycle solvent controls the overall behavior of coal liquefaction processes. The purpose of this summary is to discuss the relative merits of several chemical classes of compounds found in industrial recycle solvents. Each component will be shown to contribute in the dissolution mechanisms and should be used optimally.

The components to be considered include H-donors, phenols, H-shuttlers and H-abstractors. These will be discussed in terms of their influence on the rate and extent of coal dissolution, coal conversion, hydrogen consumption, product distribution, the chemical nature of the products, char production, and the ability to regenerate solvents.

There are presently five liquefaction processes being developed in the United States: the H-coal process by HRI; the SRC II process by DOE, EPRI, PAMCO and Gulf; SRC I by EPRI, DOE, Southern Services; SRC½ by EPRI, Conoco and Kerr-McGee; and the Exxon Donor Solvent Process (EDS). All five (under certain modes of operation) can produce clean solid fuels, which are low in sulfur and easily melted. All but the SRC I and SRC½ processes can produce substantial distillates. The SRC route with a solid-fuel product will be the focus of this discussion.

All of the above processes are extensions of technology developed in Germany during World War II. Coal is heated to 800-880°F in the presence of 2-3 volumes of a recycle solvent and 1000-5000 psi hydrogen. The products are separated and the solvent is

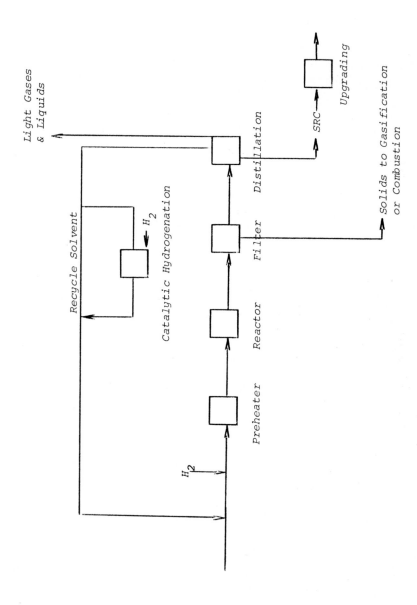

FIGURE 9-1. Present liquefaction technology.

recycled. A schematic diagram of these processes is given in
Figure 9-1.

In all cases, additional solvent-range product must be generated from the coal to make up for both physical and chemical
losses. In the SRC process, no commercial catalyst is employed
and only the intrinsic mineral matter entering with the coal acts
as a catalyst for coal liquid upgrading and/or maintenance of
proper solvent quality. An external catalyst is not necessary for
dissolution, since the coal is often substantially dissolved
through interaction with the solvent by the time the coal exits
the preheater.

H-coal is similar in operation except that a Co/Mo-Al$_2$O3
catalyst is present in the reactor, thus the product can be upgraded to higher quality than in the SRC process. The EDS process
uses a single plug-flow reactor (preheater and reactor combined)
and relies on external catalytic hydrogenation to maintain the
quality of the solvent. All 1000°F+ products are coked to gain
additional distillate materials.

The focus of this discussion is on the SRC route with a solid
fuel product, but the relationships will be relevant for other
technologies as well. A simplistic view of the SRC process is
shown in Figure 9-2. The products of the various processes are:
hydrocarbon gases, inorganic gases, soluble coal, char and unreacted coal, and additional solvent. The nature of the process

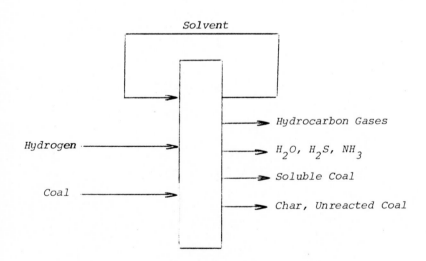

FIGURE 9-2. Selectivity in the SRC process.

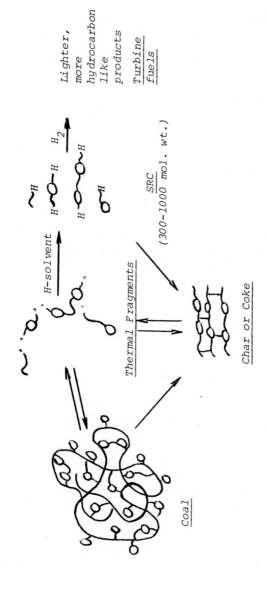

FIGURE 9-3. Chemistry involved in coal conversion.

and selectivity to the various products is primarily governed by
the composition of the recycle solvent. Similarly, changes in
the product mixture result in different compositions of the recycled solvent.

Coal conversion can be envisioned to occur in three stages:
solubilization of the coal, defunctionalization of the coal and
hydrogen-transfer, and rehydrogenation of the solvent. In each
of these stages, the nature of the solvent can affect the rates
of reaction and the distribution of the products.

Coal may be thought of as a highly crosslinked amorphous
polymer, made up of stable aggregates connected by relatively
weak crosslinks (see Figure 9-3). In the dissolution stage, because of the high temperatures, this highly crosslinked structure
thermally fragments into radicals which in the presence of
hydrogen-donor solvents are capped and appear as stable species.
These species have molecular weights in the range of 300-1000.
In the absence of hydrogen-donor solvents, the original radicals
or the smaller soluble species may recondense to form char or
coke.

The solvent governs product selectivity by controlling the
path taken by the intermediate radicals. When a bond cleaves, at
least three different pathways are available for product formation: H-abstraction, rearrangement and elimination, and addition
to aromatics. The availability of hydrogen-donors will determine
the preferred path.

Whether or not the converted coal actually physically dissolves depends on the physical solvation properties of the solvents; as indicated above, the nature of the solvent in the developing processes differs considerably. SRC solvents contain far more phenols and polyaromatics and are thus better physical solvents than H-coal or EDS solvents. This leads to some rather subtle selectivity differences. For example, as discussed in Chapter 8, components which are insoluble in a mixture will tend to self condense, forming coke at a faster rate than if they were freely dispersed in the medium. Also, high solubility aids in transporting material in and out of solid catalysts (including coal mineral matter).

Once the coal is dissolved, the process solvent controls compositional changes in the SRC and the newly formed solvent-range material, as well as the rate of gas formation.

The specific chemical properties of interest in recycle solvents and the associated chemical structures responsible for them are:

a) Hydrogen-Donor Capacity of the Solvent. - Hydrogen donors (H-donors) are believed to be important in the defunctionalization of dissolved coal and prevention of char formation. The principal sources of hydrogen appear to be partially hydrogenated aromatic hydrocarbons:

> Tetralin and its homologs, partially hydrogenated pyrene, phenanthrene, and other polycyclic aromatic compounds

b) Physical Solubilization of Coal Products. - Effective solvents for coal liquids must contain polar compounds. Assuming the concept of specific solubility parameters (9-41) applies, then good solvents should contain such components as polyaromatics, phenols, pyridines, aromatic ethers, and quinolines and their derivatives.

c) Hydrogen Transfer Capability (H-Shuttling). - Hydrogen transfer is another mechanism for dissolving coal, whereby hydrogen may be supplied from the coal itself or from the SRC to cap off radicals and form smaller soluble species. Recent reports by Neavel(9-22b) indicate that naphthalene can dissolve 80% of a vitrinite-rich bituminous coal at short contact times and at temperatures over 750°F. It was proposed that this dissolution was the result of the shuttling of hydrogen from one position in the coal to another. Naphthalene acts as an H-acceptor and the resultant free radical formed by the addition of an H-atom acts as an H-donor.   A reaction of this type is even more probable for phenanthrene or pyrene since they are better H-acceptors than naphthalene.   The structures which can contribute to good shuttling properties within recycle solvents are:

> Naphthalene and its alkyl derivatives.
> Phenanthrene and other tri- and tetraromatic hydrocarbons and their alkyl derivatives.
> Heterocyclic polyaromatics, e. g., quinoline, benzofuran, benzothiophene, etc.

d)   Chemical Structures Associated with Char Formation. - Recycle solvents may contain compounds which are prone to or which can promote char formation, in particular, heavy phenols and highly aromatic compounds.

II. CHARACTERIZATION OF RECYCLE SOLVENTS

The first step toward understanding the chemistry of recycle solvents is to determine their chemical nature.   This can be accomplished by fractionation of the solvents into discrete chemical classes, detailed characterization of the individual classes and studying reactions of whole solvents or fractions thereof to determine the changing components.   The solvent characterization work described in this chapter was conducted primarily by MRDC (Mobil Research and Development Corporation).

Gas chromatography, proton NMR and wet chemical determination of phenols have identified the major species present in process-derived recycle solvents to be:

> 70% Hydrocarbons, including condensed aromatics
> 8% Heterocyclic compounds (mostly ethers)
> 10% Monophenols
> 12% Polyphenols and basic nitrogen compounds.

The solvents used in this study were derived from Wyodak-Anderson, West Kentucky 9,14, Illinois #6 (Burning Star) and (Monterey) coals.   In some cases, several solvents were derived from the same coal under different operating conditions.

Condensed aromatics are usually the most prevalent class of compounds in recycle solvents.   In fact, a rather large fraction (30-50%) of the recycle solvents investigated is composed of just 6 compounds:

> Naphthalene
> Methylnaphthalene
> Biphenyl and/or Diphenyl Ether
> Phenanthrene-Anthracene

TABLE 9-1 . RSMC Fractions and Solvents

Fraction	Solvent	Chemical Species
RSMC-1	Petroleum ether	Saturated hydrocarbons

These are present in small concentrations and practically have no role, but in higher concentrations they could act as antisolvents.

Fraction	Solvent	Chemical Species
RSMC-2	Petroleum ether (5% benzene)	Monoaromatic hydrocarbons (including tetralin and tetralin homologs)

Their main contribution is to act as hydrogen donors; in the process, they aromatize to naphthalenes.

Fraction	Solvent	Chemical Species
RSMC-3	Petroleum ether (15% benzene)	Diaromatic hydrocarbons (naphthalene and alkyl derivatives)
RSMC-4	Tetrahydrofuran (1% ethanol)	(a) Polyaromatic hydrocarbons (fluorene, phenanthrene, anthracene)
		(b) Dibenzofuran and other furan derivatives
		(c) Nonbasic nitrogen compounds with pyrrole ring

Di- and triaromatic hydrocarbons act as good physical solvents for SRC. In addition, they act as shuttling agents in the initial stages of coal dissolution and as starting agents in the rehydrogenation step.

Fraction	Solvent	Chemical Species
RSMC-5	Tetrahydrofuran (10% water)	(a) Monophenols
		(b) Basic nitrogen compounds (pyridines, quinolines)

These are good physical solvents for coal liquids, especially short contact time SRC's.

Fraction	Solvent	Chemical Species
RSMC-6	Non-eluted	Polyfunctional compounds (e.g., hydroxyquinoline)

These compounds most likely appear in the recycle solvent due to forced distillation or as aerosols.

In attempts to further categorize the components in recycle solvents, problems have arisen with the use of gas or liquid chromatography. Gas chromatography has intrinsic limitations when characterizing solvents since classes of compounds are present with similar boiling ranges and a peak may represent a mixture of different classes. In addition, the large variation in response factors for the various chemical species in recycle solvents (even with flame ionization detection) causes quantification to be difficult. At present, 90% of the total sample can be eluted from gas chromatography columns, of which only 50-70% of the components are generally known. Gas chromatography alone is therefore not a definitive tool for understanding the composition of solvent liquids.

Several liquid chromatographic techniques have been reported which could be adapted for liquids in the boiling range of SRC solvents (9-1, 9-35-37). These methods, however, were developed to characterize specific materials other than recycle solvents from coal liquefaction. The SESC chromatographic procedure (9-1, 9-35) does not give fine separation of hydrocarbons, i. e., saturates, mono-, di- and polyaromatics. The GEC (9-36) and SARA (9-37) methods, provide good separation of hydrocarbons and functional molecules with long alkyl moieties, but are not optimal for small aromatic molecules with short chains and high functionality.

Liquid chromatographic fractionation (similar to SESC but using different solvents and alumina as the sorbent) followed by gas chromatography of the fractions has provided more relevant results. This has led to a technique, called "Recycle Solvent Multiple Characterization (RSMC)", involving more than fractionation or gas chromatography alone, since each fraction is often further examined by ^{1}H-NMR, ^{13}C-NMR, TLC, etc.

The sequence of solvents and the chemical species present in each liquid chromatography fraction by the RSMC technique, as well as the role of each class are shown in Table 9-1.

A typical preparative fractionation is shown in Figure 9-4. For rapid analysis an automated liquid chromatograph for this fractionation has been built and is described in Reference 9-34, along with the procedure for the gas chromatographic analysis.

To achieve separation of two classes a substantial difference between the relative retention factors (R_f) of the adjacent classes in the sequence must be established. The R_f values for the classes (given in Table 9-2) show that separation of RSMC-1 and 2 from RSMC-3 can be achieved by using petroleum ether, because RSMC-1 and 2 have a high R_f (about 0.8) and RSMC-3 has a low R_f in this solvent. The separation of RSMC-1 from RSMC-2 is also possible in petroleum ether since RSMC-1 has a higher average

TABLE 9-2. R_f Factors Determined by TLC for Isolated Fractions of Recycle Solvents

SOLVENTS		Pet. Ether	Pet. Ether/ 15% ΦH	THF/ 1% EtOH	THF/ 10% H_2O
FRACTIONS		1 & 2	3	4	5
RECYCLE SOLVENT	RSMC				
SN 6540	1-2	0.80			
SN 6663	1-2	0.70			
SN 6540	3	0.52	0.64		
SN 6663	3	0.48	0.65		
SN 6540	4		0.46	0.77	
SN 6663	4		0.49	0.78	
SN 6540	5			0.62	0.97
SN 6663	5			0.64	0.96

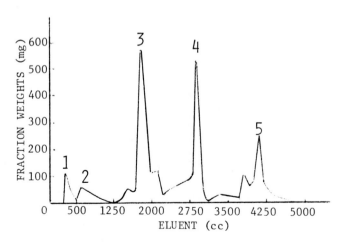

FIGURE 9-4. Preparative Liquid Chromatogram of a Process-Derived Coal Liquefaction Solvent

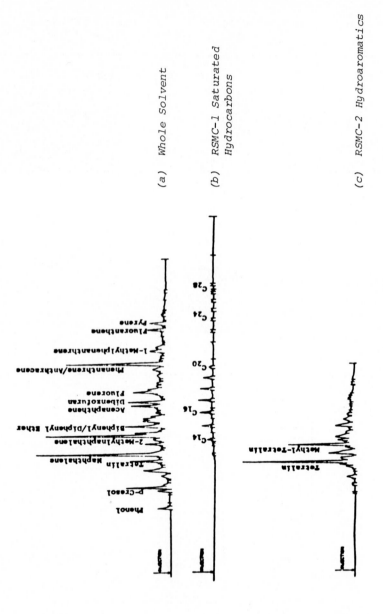

(a) Whole Solvent

(b) RSMC-1 Saturated Hydrocarbons

(c) RSMC-2 Hydroaromatics

284

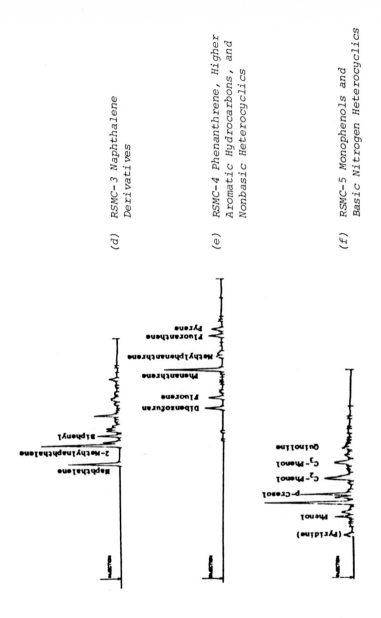

(d) RSMC-3 Naphthalene Derivatives

(e) RSMC-4 Phenanthrene, Higher Aromatic Hydrocarbons, and Nonbasic Heterocyclics

(f) RSMC-5 Monophenols and Basic Nitrogen Heterocyclics

FIGURE 9-5. Gas Chromatographic Analyses of a Whole Process Derived Recycle and its RSMC Fraction.

285

R_f than RSMC-2. This latter separation would however require long
times and an excessive volume of solvent, so another solvent mixture was inserted - petroleum ether/5% benzene - into the scheme.

Representative chemical types found in a given fraction were
identified using model compounds. It should be noted that non-
basic nitrogen compounds (pyrroles) elute in Fraction RSMC-4 and
basic nitrogen compounds elute in Fraction RSMC-5. Polyfunctional
compounds (e.g., hydroxyquinoline) do not elute.

The capability of the fractionation procedure can be shown by
comparing the gas chromatographic analyses of a whole solvent
(Figure 9-5(a)) and RSMC fractions derived from it (Figure 9-5(b)-
(f)). Key components are identified in the Figures.

A West Kentucky 9,14 recycle solvent was characterized by the
RSMC liquid chromatographic techniques before and after coal conversion. These analyses showed that the major change which
occurred was the transformation of mono-aromatic ring hydrocarbons
(e.g., tetralin) into diaromatic ring hydrocarbons. This is
shown in Table 9-3. Almost no change was observed in the content
of saturates, tri- and higher aromatic hydrocarbons, or phenols.
Gas chromatographic analyses of the phenols also showed no major
changes.

TABLE 9-3. RSMC Analysis of a Solvent Before and
After Reaction With Illinois #6 (Monterey)
(90 minutes, 800°F)

Fraction	Composition	% in Initial Solvent	% in Solvent After 90 min Reaction	Δ
RSMC-1	Saturated Hydrocarbons	8.1	6.7	(-1.4)
RSMC-2	Monoaromatic Hydrocarbons	10.9	6.5	-4.5
RSMC-3	Diaromatic Hydrocarbons	45.6	51.4	+5.8
RSMC-4	Triaromatic Hydrocarbons	22.6	22.5	—
RSMC-5	Phenols	12.7	12.0	—

III. PHYSICAL SOLVENT PROPERTIES

Both physical dissolution and chemical reactions involving hydrogen are important to a good SRC process. The chemical reactions occurring in coal dissolution involve bond-breaking and the transfer of hydrogen to coal by various mechanisms. During physical dissolution, few bonds, if any, are broken and little hydrogen is transferred from the solvent.

Several attempts have been made to generally classify the solvent properties of any solvent. Identifying the necessary solvent properties for coal dissolution helps in selecting suitable recycle solvents for any solvent. One of the first was by Hildebrand, whose solubility parameter (δ) depended on the cohesive energy of a solvent (9-42). Hildebrand, however, considered only the dispersion force (London) contribution. Later work by Hansen incorporated 3 types of contributions to the total cohesive energy: dispersion, permanent dipole-dipole and hydrogen bonding forces (9-43). A quantitative description of a solvent's ability to dissolve a given solute was later described by Teas as a three-dimensional plot. Teas later simplified the plot to a two-dimensional, three-coordinate diagram in which the three solubility parameter components were expressed as fractions of the total solubility parameter (9-39).

This approach assumes easy mixing; a negative free energy will occur between two liquids if each value of the solubility parameter components ($f_h/f_d/f_p$) for one liquid is close to the corresponding value for the other. For example, methanol with $f_h/f_d/f_p$ values of 31/23/46 is miscible with ethanol (36/19/45) (see Table 9-4). Mixtures can also approximate the behavior of a specific solvent, for example, dichlorodiethyl ether is soluble in a mixture of methanol and lignin, although individually neither is a good solvent. Thus, any liquid with solubility parameter components in ratios similar to those of SRC will dissolve it. This theory has been further supported by SRC miscibility in a mixture of solvents, which individually could not dissolve the SRC.

Looking at Figure 9-6 we can see a region of high solubility for SRC. Pyridine appears to be an excellent physical solvent for SRC although it does not extensively dissolve coal. Its physical solvent properties can be approached by mixtures containing a high proportion of aromatic hydrocarbons (with dispersion forces) and a moderate amount of polar and aliphatic compounds. The plot also shows that excess hydrogenation of the solvent during regeneration may cause the SRC to become insoluble. Typical equilibrium recycle solvents contain large proportions of aromatic hydrocarbons and phenols necessary for good physical

TABLE 9-4. *Solubility Parameters and Fractional Parameters of Organic Liquids (Reference 9-42).*

	f_d	f_p	f_h		f_d	f_p	f_h
Methanol	31	23	46	Pyridine	56	22	22
p-Cresol	49	17	34	Chloroform	67	10	23
Dimethyl sulfoxide	37	33	30	Benzene	76	7	17
Acetone	50	37	13	Tetralin	83	4	13
Tetrahydrofuran	55	22	23	Hexane	95	2	3
Aniline	55	21	24	Water	19	22	59
Ethanolamine	31	32	37	Acetic acid	40	19	41
Dimethyl formamide	41	32	27				

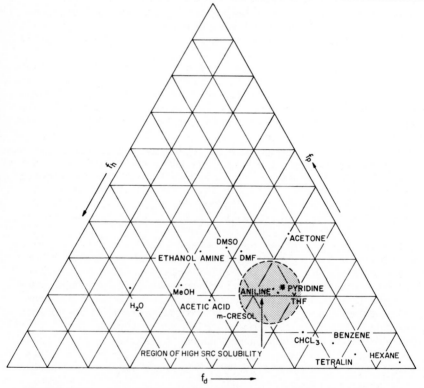

FIGURE 9-6. *Fractional Solubility Parameters for Common SRC Solvents.*

solubilization. Polyaromatic components tend to have solubility
parameters more closely approaching pyridine than do simple
aromatic compounds and are therefore beneficial.

Being a good physical solvent for SRC, however, is only one
criterion in the formulation of a good recycle solvent. Pyridine
in the absence of gaseous hydrogen will not dissolve coal at
either long or short contact times although it is one of the best
physical solvents for SRC. Thus the chemistry of the solvent is
even more important.

IV. DISSOLUTION MECHANISMS

A. Hydrogen Donor Effects

It is generally agreed that H-donors are "good" components in
recycle solvents. But just what is an H-donor? By definition,
an H-donor is a species which can give up one or more hydrogens
to another species when they interact. The critical factor in
hydrogen donation is the demand for hydrogen as dictated by the
other components of the reaction mixture and the process conditions.

In the SRC process, there are at least five sources of hydrogen (9-1 through 9-4).

>   Hydrogen gas
>   Solvent
>   Coal
>   SRC
>   Residue

The latter four contribute their hydrogen through the following chemical transformations:

>   Hydroaromatic dehydrogenation
>   Aromatic substitution
>   Char formation

The particular mechanism by which hydrogen is donated depends
on the demand. Consider, for example, the reactions of benzophenone (an H-acceptor) with the synthetic solvent components.
The experiments were non-competitive such that benzophenone
abstracted hydrogen from the only available source in each case.
Even non-hydroaromatic components of the synthetic solvent acted
as H-donors. We have indications that benzophenone can even
react thermally with elemental hydrogen at liquefaction temperatures (800°F).

	CH₃ structure	Tetralin	p-Cresol
	2-Methylnaphthalene	Tetralin	p-Cresol

% Benzo-
phenone
Conversion
850°F, 1 hr   4   16   16

But given a choice, hydrogen donation by the most active
H-donors is preferred. This is shown in Figure 9-7(a), where a
synthetic solvent containing some additional 9,10-dihydrophenan-
threne was reacted with benzophenone. It can be clearly seen
that as the benzophenone is converted to diphenylmethane, the
conventional hydroaromatic H-donors are the first to provide
hydrogen. In fact, there is some evidence of a sequential nature
to this donation even among hydroaromatics. As the hydroaromatics
are depleted, p-cresol and even methylnaphthalene become major
sources of hydrogen.

In some ways benzophenone is less demanding than coal as we
believe that the rate-determining step in the above experiments
was the slow steady generation of a reactive intermediate from
benzophenone.

As will be shown in the section on hydrogen consumption, the
coal, not the solvent, dictates the amount of hydrogen that must
be donated during liquefaction. Thermal reactions generate a
continual demand for hydrogen, which will be taken from any
available source.

In the case of coal, the demand for hydrogen is relatively
high during the initial stages of the reaction due to the massive
thermal fragmentation of labile components of coal. As the
conversion proceeds, the remaining bonds that can be thermally
cleaved are stronger and create a lower rate of demand.

Figure 9-7(b) presents the results of a reaction between
Illinois #6 (Monterey) coal and a synthetic solvent similar to
that described above for the reaction with benzophenone. It can
be seen that again there is a preference for donation of hydrogen
from hydrophenanthrene over tetralin. In this particular case,
the molar reactivities were 40/1, respectively. Little evidence
for p-cresol or methylnaphthalene participation in the H-donation

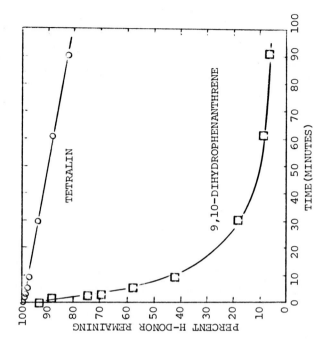

FIGURE 9-7(b). Relative Reactivity
of H-Donors Toward Coal.

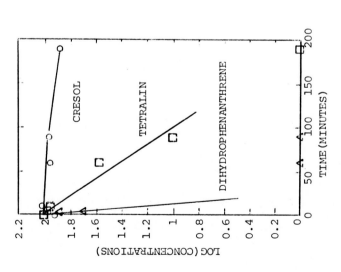

FIGURE 9-7(a). Relative Reactivity
of H-Donors Toward Benzophenone.

D. Duayne Whitehurst *et al.*

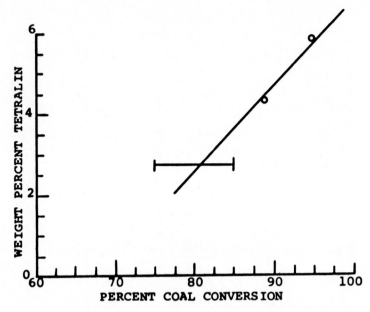

(a) *Wilsonville, Illinois #6 Coal*

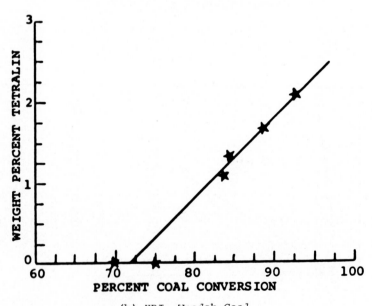

(b) *HRI, Wyodak Coal*

FIGURE 9-8. *Conversion vs. Wt.% Tetralin in Process-Derived Solvents*

was observed in this reaction; furthermore, the tetralin was not completely consumed.

What are the H-donors in a process-derived recycle solvent; what are their concentrations; and how do they control the selectivity of product formation during SRC processes? For low boiling range solvents, the concentration of tetralin can be used as an indicator of the hydroaromatic concentration in the solvent and this correlates with coal "conversion" in pilot operations (9-1). This is illustrated in Figure 9-8. In the case of severe, long-time conversions such as these, one cannot really distinguish between the true conversion of coal and the possibility of char formation which would be accounted for as unconverted coal.

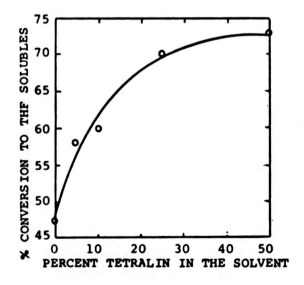

FIGURE 9-9(a).   Conversion of Indiana V Coal to THF Solubles.

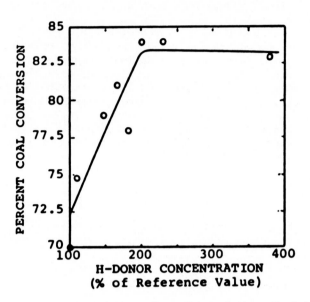

FIGURE 9-9(b).   Conversion of Wyodak-Anderson Coal to
Pyridine Solubles.

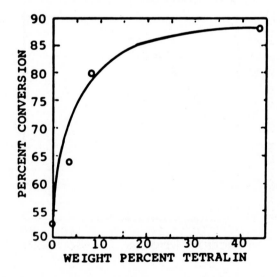

FIGURE 9-9(c).   Effect of H-donor on Wyodak-Anderson Coal
Conversion.

In expressing the concentration of H-donors in a process-derived solvent, it is common practice to relate the donor activity of a particular solvent to one observed for a less complex solvent having a known concentration of a specific H-donor. Tetralin is commonly chosen as the representative hydroaromatic H-donor (9-4).

Figure 9-9(a) shows the conversion to THF solubles of Indiana V coal versus tetralin concentration in a solvent activity test developed by Conoco personnel and used routinely at the Wilsonville PDU (9-6). Figure 9-9(b) shows the conversion of Wyodak-Anderson coal to pyridine-solubles as a function of the H-donor concentration of the solvent (expressed as percent of a reference value) as defined by EXXON (9-5(c)). Figure 9-9(c) shows MRDC results for the conversion of Wyodak-Anderson coal to pyridine solubles as a function of the tetralin concentration in a synthetic solvent.

TABLE 9-5. *H-Donor and Phenol Contents of Various Solvents*

	Wt% Hydro-aromatics	"Available" H g/100g	Wt% Phenols	-OH meq/g
Exxon[a] (patents)	30-50	.8-3 1.2 preferred	–	.08-.62[b]
SRC-I	8-12	.2-.6[c]	20-30	1.0-2.4
Wilsonville PDU				
SS*40[d]	40	1.2	18	1.7
SS*8	8	.24	18	1.7
SS*4	4	.12	18	1.7
SS*0	0	0	18	1.7

[a]*References 9-5(d)-(f).*

[b]*Reference 9-7.*

[c]*About one-half of this value is represented as hydroaromatic hydrocarbons.*

[d]*This designates one of our synthetic solvents in which the tetralin concentration is given after the asterisk.*

Each example shows that conversion is responsive to the concentration of hydroaromatic H-donors up to a particular concentration level beyond which further H-donor addition has little beneficial effect.

Exxon researchers have also shown that at low levels of H-donor, elemental hydrogen is synergistic for the conversion of coal and a lower quality solvent can be used. At high H-donor levels, however, elemental hydrogen shows little added advantage (9-1).

To determine the effects of different concentrations and types of H-donors on coal conversion and product selectivity, a series of conversions was conducted with Illinois #6 (Monterey) bituminous coal and Wyodak-Anderson subbituminous coal at 800°F, 90 minutes, and 1000 psi H_2. The compositions of the solvents used in this series compared with those of two process-derived recycle solvents are shown in Table 9-5.

1. Coal Conversion

The extent of coal conversion at long times, is responsive to the hydroaromatic content of recycle solvents. In this series of synthetic solvent conversions, clear trends were observed of increased coal conversion with increased concentration of tetralin in the solvent as shown in Figure 9-10. It is interesting to

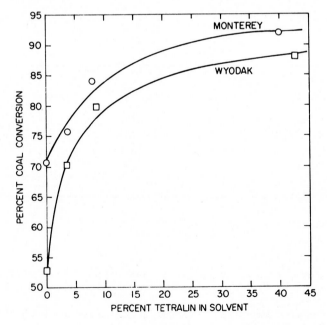

FIGURE 9-10. *Coal Conversion vs. Tetralin Concentration.*

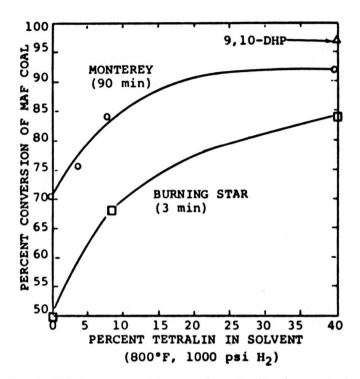

FIGURE 9-11(a). Rate of Conversion of Bituminous Coal.

note, however, that even with no classical H-donor (hydroaromatic)
in the solvent, 70% of the bituminous coal Illinois #6 (Monterey)
was converted. The intrinsic extractability of this coal was
only 20%; thus a substantial increase in conversion was achieved
at long time without any hydroaromatics being present.

Long time "conversions" can be confused by the re-formation
of solids (e.g., char) from soluble coal products, and this may
have happened to some extent at low hydroaromatic concentration.
At short times, char formation is not considered a major problem;
thus, coal conversions at short time with varying H-donor contents would not be affected by this factor. Figure 9-11(a) shows
that even at short times, in which there is relatively little
H consumption, the conversion of Illinois #6 (Burning Star) coal
is proportional to the concentration of hydroaromatics. This
means that the rate of coal dissolution is dependent on the
interaction of H-donors with insoluble intermediates and that the

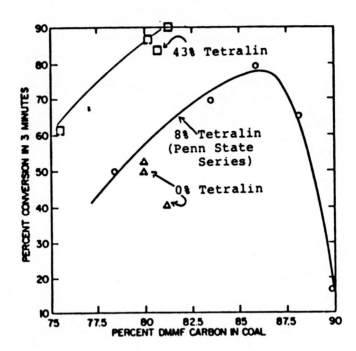

FIGURE 9-11(b).   3 Minute Coal Conversions vs. DMMF Carbon
at Various Tetralin Levels.

thermal fragmentation of bonds is not the only controlling factor.
This coal gives very similar performance to the Illinois #6
(Monterey) coal with our synthetic solvents and can thus be com-
pared to the results above.

The effect of increased H-donor concentration leading to
increased coal conversion, regardless of the rank of the coal can
be seen in Figure 9-11(b) where coal conversions at short times
(3 minutes) are shown. The conversions with the H-donor-limited
solvent (8% tetralin) are appreciably lower than are those with
the H-donor-rich solvent (43% tetralin) if compared at the same
rank.

It is significant that even at very short time (about 3 minutes) hydrogen gas satisfied a large portion of the hydrogen demand created by coal radicals when in H-donor-limited solvents (8% tetralin). Figure 9-12 shows that large amounts of hydrogen are consumed from gas-phase hydrogen by low-rank coals. With higher-rank coals (80% C) the hydrogen consumption is much lower.

The effects of temperature were discussed in Chapter 7; the interrelationships of temperature and donor concentration can be seen from the data below. Although raising the temperature or lowering the donor concentration lower the SRC yield at long times, it can be seen in Figure 9-13 for a bituminous coal that raising the temperature results in moderate increases in both residue and light product yields, while lowering the donor at either high or low temperature results in a substantial increase in residue and a moderate decrease in light product yields. The effects are similar for subbituminous coal, which is more sensitive to tetralin concentration.

SRC Yields With Various Donor Levels and Temperature

| | Illinois #6 (Monterey) SRC | | | | | |
	800°F				860°F	
Tetralin Level in Solvent (%)	40	8.5	4	0	0	40
SRC Analysis[a] (%)						
C	85.08	84.75	85.72	85.10	83.92	86.72
H	5.92	5.50	5.57	5.47	4.92	5.75
O	5.83	6.25	5.41	5.80	8.29	4.66
N	1.67	1.56	1.46	1.62	1.20	1.88
S	1.51	1.94	1.84	2.02	1.66	0.99
Ash	0.00	0.36	1.14	0.78	0.98	0.35
Cl	0.00	0.00	0.00	0.00	0.00	0.00
H/C	0.83	0.77	0.77	0.77	0.70	0.79
Percent of Coal	62.36	58.00	64.00	56.00	48.22	37.89

[a]*CHONS given on MAF basis.*

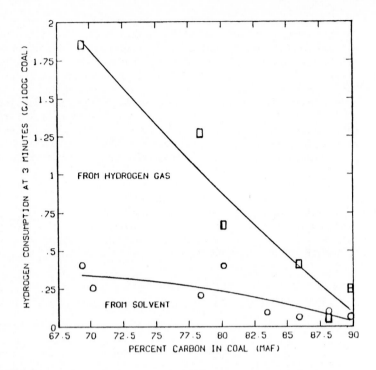

FIGURE 9-12.   Hydrogen Consumption from Solvent and Gas vs. MAF Carbon.

A speculation on the mechanisms that may be involved which accounts for the response of conversion at short times to H-donor content is shown in Figure 9-14. For the sake of discussion, ethylene bridges are used to illustrate the sequence of reactions involved. Other linkages are also possible, e.g., benzylic ethers. In this mechanism it is assumed that the first step in the conversion is thermal bond rupture (independent of H-donor) but that more than one bond must be cleaved before solubility is achieved. If two bond ruptures must occur, then the probability of reversible bond re-formation is high since it is very unlikely that both bonds cleave simultaneously. Thus, interaction of an H-donor with the radicals formed upon the first bond rupture must occur before the second bond is cleaved and the product becomes soluble. The kinetics of such a process would be consistent with the rate dependence on the concentration of H-donor.

Our calculations on the number of H atoms added per average molecule at short times are also consistent with the mechanism.

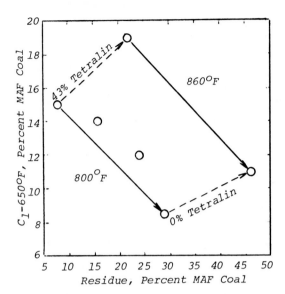

FIGURE 9-13.   Light Products vs. Residue

The figure also illustrates that in the event sufficient H-donors
are not available, the radicals may add to nearby aromatic rings
and can produce linkages that are more stable (as discussed
earlier in Chapter 7).   Implicit in this mechanism is the assump-
tion that hydroaromatics are intrinsically more reactive than H
gas; thus their concentrations are very important.

   2.   Product Distribution and Quality

   The solvent's composition controls the product distribution
in coal conversions.   The primary goal of the SRC process is to
produce a low sulfur low-ash vacuum residue, which is also easily
melted.
   SRC is defined in this book as the pyridine-soluble product
that boils above 650°F (equivalent atmospheric distillation).
Since the synthetic solvents used in these experiments were composed of low-boiling materials, separating the solvent from the
SRC usually was not difficult.   However, for this series of reactions, the hydroaromatic H-donor content was systematically
lowered.   In addition, as the hydroaromatic level decreased, the

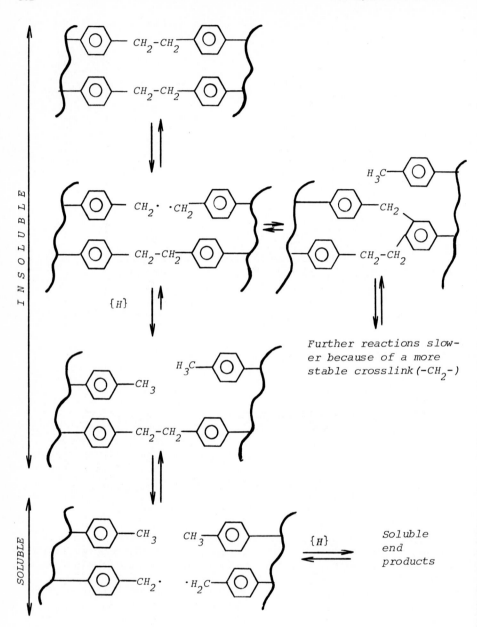

FIGURE 9-14. *Speculative mechanism to explain the dependence of coal dissolution on H-donor concentration at short times.*

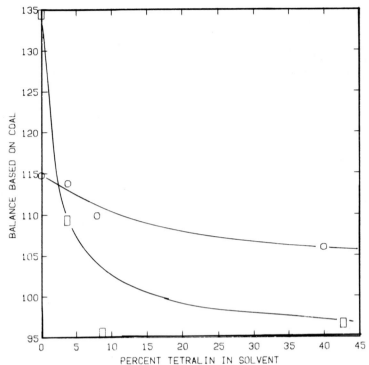

FIGURE 9-15. *Balance vs. Tetralin Concentration.*

□ *Wyodak*
O *Monterey*

"balance", calculated from the appearance of products (supposedly from coal) that were not originally present in the mixture became extremely high (see Figure 9-15).

Solvent reactions were then investigated and several solvent reactions were observed to increase substantially as the tetralin level in the synthetic solvent mixture was decreased from 43% to 0%; some contributed to the high "balance".

Most of the reactions detailed below occurred to a negligible extent when the solvent contained 43% tetralin; however, as the tetralin content was reduced, the conversions increased. The reactions involved:

(1) the disappearance of p-cresol. The amount was accounted for primarily by the methyl transfer reaction in which p-cresol disproportionated to phenol, ethyl and dimethyl phenols. Some losses of p-cresol due to condensation reactions with other

solvent species, coal and coal products were also noted. The
mechanism for these reactions is discussed in detail later in this
chapter.

(2) 2-methyl-naphthalene also underwent significant condensation reactions as well as methyl transfer reactions similar to
those of p-cresol. The major products identified as originating
from 2-methyl-naphthalene were dimers. Some demethylation to
naphthalene was also observed.

etc.

(3) Cross condensation products of p-cresol and 2-
methylnaphthalene were also found. These compounds appeared in
the SRC yields.

Positive indentification of these products was achieved by
GC-mass spectroscopy. Surprisingly, the mass spectra showed that
in most cases the dimers (trimers, etc.) were fused-ring compounds.
Most dimers were substituted methyl aromatics. In many cases the
detected compounds were hydroaromatic analogs of these fused-
ring systems. Furthermore, the dimers involving p-cresol were
usually not phenols, because when p-cresol condenses to form a
fused-ring dimer, the resulting phenol readily dehydroxylates.

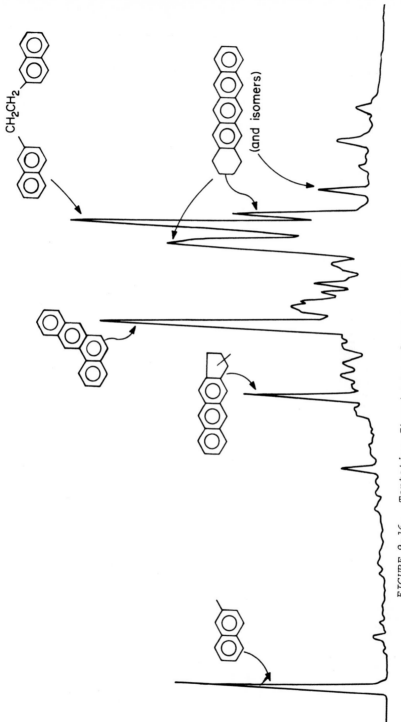

CH₂CH₂

(and isomers)

FIGURE 9-16. Tentative Structures of Solvent Dimers From AC-159.

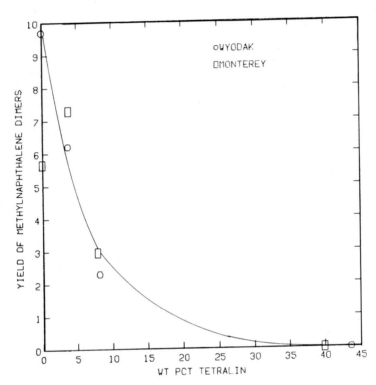

FIGURE 9-17.   H-Donors and Yield of Methylnaphthalene Dimers.

A rather interesting observation was that independent of the coal being converted or the level of hydroaromatics, the distribution of the dimeric products was always about the same. The major single dimeric product was 1,2-dinaphthylethane. Figure 9-16 shows a typical gas chromatograph of the SRC range products which were derived from the solvent.

Increasing the tetralin concentration dramatically reduces the yield of dimeric products; this is illustrated in Figure 9-17. An explanation for this phenomenon will be presented later.

After correcting for the above solvent products, product balances were much more reasonable, although they still appeared high when no hydroaromatics were in the solvent.

3. Hydrogen Consumption

The high product yield from the solvent described above is believed to be due to the fact that fragmentation of the reactive

TABLE 9-6. *Availability of Hydrogen From Various
Sources at Start of Liquefaction*

800°F, 1000 psi H_2, 5/1 solvent/coal
(all data are reported as g H/100 g coal)

Approximate % Tetralin	40	8.5	4	0
g H in Hydroaromatics	6.1	1.3	.6	0
g H in Aromatic methyls	2.3	3.4	3.6	3.7
g H in Gas[a]	.5	.5	.5	.5
g H in Coal[b]	2.0	2.0	2.0	2.0
g H - Total	10.9	7.2	6.7	6.2

[a]This includes only the H_2 soluble in the medium. Assumed to be
equivalent to that in cresote oil. Reference 9-7. More H_2 is
available from the gas phase if the soluble H_2 reacts.

[b]Assumes 37% of H in coal can be removed by dehydrogenation
(Reference 9-3b).

bonds in coal is independent of the solvent composition or the
presence of molecular hydrogen. The demand for hydrogen created
by the thermally-produced radicals must be satisfied, and in the
absence of hydroaromatic H-donors other sources are called upon.
    Accordingly, in the absence of hydroaromatics, aromatic methyl
groups become major sources of hydrogen (donating one hydrogen/
molecule). On doing so they produce benzylic radicals which stabilize by several routes. They can dimerize producing ethylene
linkages or they can add to aromatic rings producing methylene
linkages and releasing one more hydrogen. The ethylene linkages
are potentially regenerable as "H-donors" as they can be thermally
or catalytically cleaved and restored to aromatic methyl groups.
In some instances when two methylenes link aromatics onto adjacent
aromatic carbons, fused aromatics can result. Hydroaromatic
structures and methyl groups of coal and SRC also become sources
of hydrogen if the solvent cannot supply it. The concentrations
of various types of hydrogen have been calculated for each synthetic solvent composition. These data are presented in Table
9-6.
    These calculations assume that only the molecular hydrogen
which is dissolved is instantaneously available to the radicals and

that the solubility is the same as that in cresote oil (9-7). It
is also assumed that 37% of the hydrogen in the coal can be removed by dehydrogenation. This value comes from catalytic dehydrogenation studies reported for coals similar in rank to those
of this study (9-3(b)).

The data in the table show that the coal is potentially a
major source of hydrogen, equivalent to a concentration of 13%
tetralin at a 5/1 solvent/coal ratio.

In the course of coal liquefaction, hydrogen is needed to
stabilize radicals, reject heteroatoms, and aid in lowering the
molecular weight of coal products to distillate liquids and gases.
For the purpose of the following discussion, we shall assume that
the requirement for hydrogen in our studies was for heteroatom
removal, distillate production and gas production. This assump-
tion appears reasonably correct since the combined yields of SRC
and residue have been shown to represent 75 to 85% of all of the
coal products in a series of runs under standard conditions (9-34).

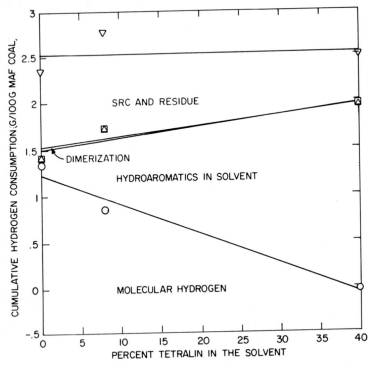

FIGURE 9-18(a). *Hydrogen Consumption with Illinois #6*
(Monterey) Coal.

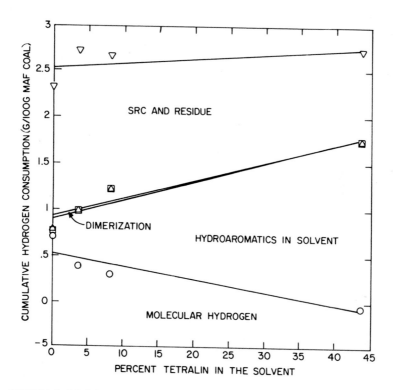

FIGURE 9-18(b).   Hydrogen Consumption with Wyodak-Anderson
Coal.

The consumption of hydrogen from coal (i.e., SRC + residue)
can thus be calculated from the combined yield of hydrogen in
these products by comparing this value to the original weight of
hydrogen in the coal.   Figures 9-18(a) and 9-18(b) show the
amounts of hydrogen donated from various sources as the concentration of hydroaromatics in the solvent was changed.   The lines
shown on these figures represent first order fits of the data
points and are thus for illustrative purposes only.
    First it should be noted that the total hydrogen consumption
for each coal was relatively independent of the hydroaromatic
concentration in the solvent.   In the range of 10 to 20% tetralin
in the solvent, the SRC and residue provide as much hydrogen as
the hydroaromatics for heteroatom removal and the production of
lighter products.

Hydrogen consumption from molecular hydrogen was much more significant for Illinois #6 (Monterey) coal than for Wyodak-Anderson coal, presumably because of catalysis by mineral matter (see Chapter 6). With both coals, as the hydroaromatics in the solvent became limited, consumption of molecular hydrogen increased. Hydrogen consumption from the dimerization of aromatic methyl groups in the solvent was a minor factor.

Hydrogen consumption from the SRC can also be observed by examining the composition of SRCs produced in various ways. For example, in Table 9-7 a general trend of lower SRC and residue H/C ratios with lower hydroaromatic concentration can be seen. With Illinois #6 (Monterey) coal, little variation was noted in the aromatic content of the SRC; however, with Wyodak-Anderson coal a clear trend was found for increasing aromatic content in the SRC as the hydroaromatic concentration of the solvent was lowered.

Table 9-8 presents some additional data as well as the result of another coal conversion in which the concentration of labile hydroaromatic "H" was the same as the highest tetralin concentration (1.2 g H/100 g solvent) but the nature of the hydroaromatic was changed (9,10-dihydrophenanthrene was included). In this case not only was the H/C ratio of the SRC improved but the overall coal conversion was higher and the degree of desulfurization and deoxygenation was increased. It can also be seen that the sulfur content of the SRC decreases with increasing hydroaromatic quantity and quality in the solvent. It should be recalled that 9,10-dihydrophenanthrene was 40 times more active than tetralin. Thus, not only is the concentration of hydroaromatics important but the lability of the hydroaromatic is also critical.

4. SESC Fraction Yields

By using liquid chromatographic fractionation, the changes in SRC composition due to altered hydroaromatic content in the solvent can be observed.

The yield distributions of the different chemical classes found in these SRC's are shown in Figure 9-19(a) and (b). With Illinois #6 (Monterey) coal, little specific change in composition was noted other than an increase in the monophenol content at low hydroaromatic content in the solvent. We believe this was due to condensation of p-cresol with coal products, with retention of the phenolic -OH. (This point will be discussed in more detail later in Section C. The Effect of Phenols.)

With Wyodak-Anderson coal, however, a marked change in the distribution of the various chemical classes found in the SRC occurred as the hydroaromatic content was lowered to less than 8% tetralin. The major change was a selective loss in the highly

TABLE 9-7. Analyses of SRC's and Residues for Conversions
with H-Donor Variation

COAL	Illinois #6 (Monterey)				Wyodak-Anderson			
% Tetralin in Solvent	39.9	8.0	3.8	0	43.7	8.2	3.7	0
%Ar$_C$ in SRC	79	82	79	82	74	82	85	87
H/C of SRC	.83	.78	.79	.78	.89	.82	.80	.81
H/C of Residue	.75	.53	.58	.55	.59	.51	.53	.55

TABLE 9-8 . Conversion and SRC Composition with Various H-Donors
(Illinois #6 (Monterey Coal), 90 minutes)

	Tetralin				Tetralin and 9,10-dihydro-phenanthrene	Coal
Available meq H/100g	0.00	0.12	0.24	1.2	.4 and .74	
Conversion	70.7	75.5	84.1	92.0	96.9	
H/C	0.78	0.79	0.78	0.83	0.86	.86
Pct. H	5.5	5.6	5.4	5.9	6.1	5.58
Pct. S	1.8	1.7	2.1	1.5	1.3	

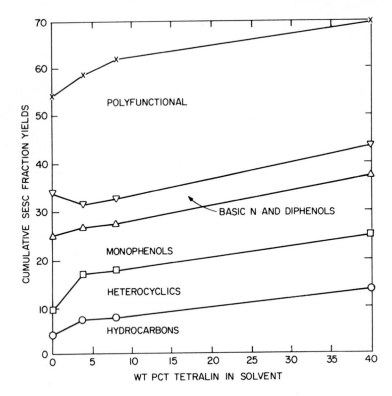

FIGURE 9-19(a). Illinois #6 (Monterey).

functional and monophenolic components of the SRC. The loss of
these components closely parallels the major decrease in SRC
yield and loss in "conversion" at these low levels of hydroaromatics. Thus the fractions that are misssing from SRC's
produced at low tetralin concentration are those that are most
prone to condense to form char when they are in solution. As
was shown in Chapter 7, in these series of runs, these materials
most likely never became pyridine-soluble; if they were initially
dissolved, however, they may have recondensed.

In Figure 9-20(a) and (b) the total cumulative hydrogen
inventory of the coal products is shown as a function of the
initial percent tetralin in the solvent. It can be seen that in
the lowest donor solvents, there is a significant increase in the
amount of hydrogen that appears in the residue of the final
products. This is more than just a reflection of greater residue
yield, as condensed solvent components also increase hydrogen
content of the residues.

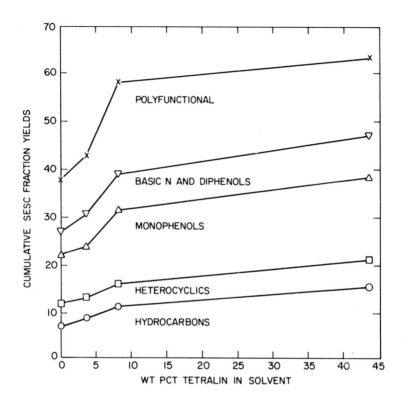

(b). Wyodak-Anderson.

FIGURE 9-19. Effect of H-donor on SESC Fraction Yields.

With both coals the major portion of the hydrogen in coal products is found in the SRC. Also, the hydrogen associated with light gases was about 20% of the total hydrogen in the products, which represents a non-beneficial consumption of hydrogen.

The yield of gas was not grossly different in reactions with solvents of varying hydroaromatic content while the yield of SRC was significantly lowered. Thus, the efficiency of hydrogen utilization appears to be poorer when solvents of low hydro-aromatic content are used. One subtle relationship between the hydroaromatic content of the solvent and the efficiency of hydrogen utilization has been reported by Exxon (9-5(c)). They have shown that, at comparable times of reaction, the yields of C_4-1000°F products from coal are lower for solvents of low

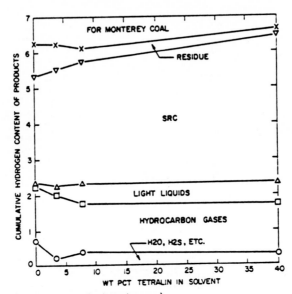

(a)   *Illinois No. 6 (Monterey).*

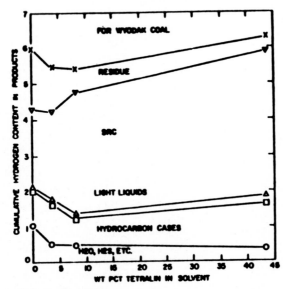

(b)   *Wyodak-Anderson.*

FIGURE 9-20. Effect of H-Donor on Hydrogen Content of Products

TABLE 9-9 . Effect of Residence Time on Yields
Using Low SQI Solvent[a]

Yield Periods	378,379,393 394,395,404 405,406,407 408	485,486 487,488
Residence Time, min.	25	60
Solvent Quality Index	4.5	4.7
Overall Yields, lbs/100 lbs Dry Coal		
C_1-C_3	5.9	8.6
C_4-1000°F Liquid	26.2	24.2
Conversion to 1000°F⁻, lbs/100 lbs Dry Coal	47.7	50.2
Conversion to Pyridine Solubles, %	83.4	84.5

[a] Reference 9-5 (c) .

Solvent Quality Index (a proprietary measure of the hydrogen donor
content of the solvent). If the same yield of C_4-1000°F is to be
achieved with a hydrogen donor-deficient solvent, longer times of
reaction must be used and the consequence is that more gases are
produced. This point is illustrated in Table 9-9 (9-5(c)).

B.   The Effect of Polyaromatic Hydrocarbons (Hydrogen-Shuttlers)

     Although H-donors are major controlling species in coal
liquefaction, they are not the sole controlling factor. Several
workers (9-3,9-4,9-8,9-9) have reported that coal or SRC can provide significant amounts of hydrogen in liquefaction processes.
In fact, it has been reported (9-8,9-9) that at relatively short
times of reaction over 80% of the reactive macerals of coal
(vitrinite) can be converted to pyridine soluble form in the
absence of hydrogen gas or a hydroaromatic but in the presence of
a polyaromatic non-donor solvent (e.g., naphthalene) which could
aid in the redistribution of hydrogen among different coal species.
This phenomenon was termed H-shuttling.

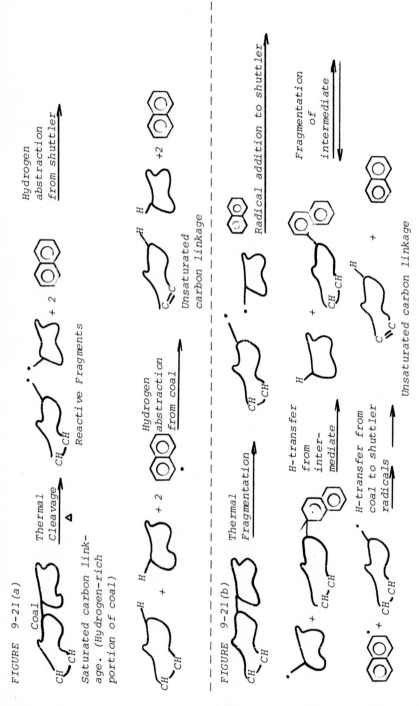

FIGURE 9-21(a)

Coal — Thermal Cleavage → CH CH Reactive Fragments

Saturated carbon linkage. (Hydrogen-rich portion of coal)

+ 2 — Hydrogen abstraction from shuttler →

+ 2 — Hydrogen abstraction from coal →

C=C Unsaturated carbon linkage

FIGURE 9-21(b)

CH CH — Thermal Fragmentation →

H-transfer from intermediate →

H-transfer from coal to shuttler radicals →

CH CH — Radical addition to shuttler →

Fragmentation of intermediate

CH CH + C=C Unsaturated carbon linkage

FIGURE 9-21(c).

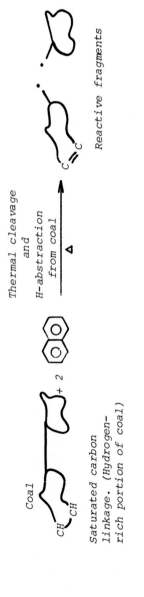

Coal

Saturated carbon
linkage. (Hydrogen-
rich portion of coal)

Thermal cleavage
and
H-abstraction
from coal

Reactive fragments

H-re-addition
to coal radicals

Unsaturated carbon
linkage

FIGURE 9-21(a-c). H-shuttling mechanisms.

Several possible mechanisms by which shuttling may occur are shown in Figures 9-21(a) and (b). For these mechanisms to become effective, the polyaromatic solvent must be thermally stable, yet labile to either hydrogen abstraction, radical addition, or to hydrogen addition.

In such sequences no net change would occur in the shuttler or the H/C ratio of the coal; however, the content of unsaturated carbons in the coal products would increase. Because no net hydrogen is donated, there is an intrinsic limit on the extent to which such a reaction is useful. The quantity of labile hydrogen available from the coal itself is limited and the beneficial effect of this process is only short-lived. The amount of available hydrogen of this nature can amount to as much as 1.9 g H/100 g MAF coal for bituminous coals. This was demonstrated by Wender et al. by the catalytic dehydrogenation of coal to produce hydrogen, which is a measure of the hydroaromatics content of coal (9-3). If the soluble coal is further thermally fragmented the demand for hydrogen exceeds the coal's capacity of labile hydrogen; it is then abstracted at the expense of char formation.

All of the proposed mechanisms shown in Figure 9-21 assume that the initial step of the reaction is thermal fragmentation of the coal to produce free radical species. Even unheated coals are known to contain significant quantities of free radicals and on heating to 800°F the quantity of free radicals greatly increases. This increase in free radical concentration has been shown to be associated with the reactive macerals in coals (e.g., exinite and vitrinite) (9-11) while unreactive macerals (fusain) do not increase in radical concentration on heating. This is shown in Figure 9-22 for a bituminous coal which was heated for one hour at the indicated temperature, then cooled and the electron spin resonance spectra determined. After cooling to room temperature, the vitrain sample still had 20 mmoles radical/100 g coal. It is interesting to compare this amount with the amount of hydrogen required to stabilize short contact time SRC. Our estimates (9-2) as well as those of others (9-21, 9-22) indicate that about 0.3% hydrogen consumption occurs in converting 80% of bituminous coal to pyridine-soluble material in 10 minutes. This amounts to 300 mmoles/100 g. For the coal which had been heated and cooled but not capped, about 7% of these 300 mmoles of radicals could still be observed at room temperature.

In a similar set of experiments, Wyodak-Anderson coal retained 17.4 mmoles/100 g free radicals after being heated for 2 hours at 840°F in naphthalene/N_2 cooled and analyzed by ESR. When the same coal was heated in tetralin/H_2, only 4.2 mmoles/100 g of free radicals were observed after cooling to room temperature (9-20).

Thus, on heating coals to liquefaction temperatures, high concentrations of free radicals are formed by thermal bond breakage. If conventional hydroaromatic H-donors are present, the radicals are stabilized by hydrogen transfer from the solvent. If no hydrogen donors are present, the radicals must stabilize in other ways (e.g., by recombination, aromatic ring alkylation, or H-abstraction from the coal itself). H-shuttlers can aid in this stabilization through the mechanisms shown in Figure 9-21.

A number of authors have reported on the ability of non-donor polyaromatic solvents to "dissolve" bituminous coals (9-22 through 9-25). Coal dissolution by these solvents is generally associated with major hydrogen exchange reactions between coal and the solvent. The extent of H-exchange by phenanthrene with coal at close to liquefaction conditions (662°F, long times) was measured using deuterium, tritium, and ^{14}C tracers (9-24(a)). That work showed that 9-15% of the hydrogen of the coal exchanged and 80-80% of the coal dissolved. Some phenanthrene did, however, chemically condense with coal products (probably by a mechanism similar to that of Figure 9-21(b)). Even when tetralin was used as a solvent for coal, the extent of exchange of aromatic hydrogens was about the same as the extent of H-donation by the tetralin (9-24). Hydrogen exchange between coal and tetralin also occurred in the saturated ring (predominantly in the beta position to the aromatic ring) which indicates that hydrogen abstraction in that position does not necessarily lead to rapid naphthalene production.

A recent article showed that extract yields, after pyrene treatment at 700°F, depended on the rank and fluidity of the coal. Maximum yields were found between 83 and 88% MAF C; coals with high fluidity gave high yields (9-44).

The relative ability of a series of pure polyaromatic compounds to dissolve a bituminous coal has recently been reported by workers at the National Coal Board, Stoke Orchard, Great Britain (9-25). In the reported studies, a coal containing 84% DAF carbon (International Classification 602) was heated to 752°F for 1 hour in the presence of a 3/1 weight ratio of various compounds under an inert atmosphere. Conversions were measured by quinoline solubility of the products. A summary of their results is presented in Table 9-10.

Several points can be made from these data. With the exception of anthracene, the more highly condensed ring systems were more effective in promoting coal conversions. This result is in line with an increasing efficiency of coal dissolution with increasing boiling point of the solvent as noted by those authors and is also the experience of operating SRC PDU's.

The higher conversions noted for methylated naphthalenes and fluorene relative to totally aromatic molecules is in agreement

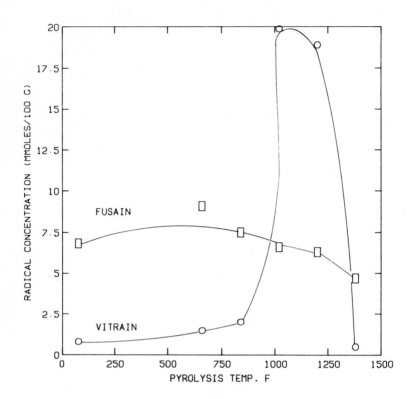

FIGURE 9-22.   Free Spin Concentration of Vitrain and Fusain for Redstone Seam Coal (Buckhannen, W.Va.)

with previous discussions on the ability of aromatic methyl (or other benzylic hydrogens) to donate hydrogen to coal radicals. The high conversion with carbazole solvent may also be due to this mechanism, but other factors may be equally important (such as activation of the aromatic ring by nitrogen substitution).

Figure 9-23 shows the National Coal Board's results for coal dissolution in various boiling cuts of raw anthracene oil and hydrogenated anthracene oil (9-25). With the exception of fractions which contain acenaphthene, a hydrogen donor, there is a clear trend of increasing conversion with increasing boiling point. This reflects the sensitivity of conversion to the size of polyaromatics in the solvent.

The hydrogenated solvents do not show this trend, because the hydroaromatics alone are effective in producing high yields of products (80%). H-donors and H-shuttlers can work together

TABLE 9-10. *Coal Dissolution in Pure Polyaromatic Non-Donor Solvents*[a]
(3/1 polyaromatic/coal, 752°F, 1 hour, inert atmosphere)

Polyaromatic Solvent	% Standard Extraction Efficiency[b]
(biphenyl)	13
(naphthalene)	14
(anthracene)	32
(methylnaphthalene, CH₃ position)	48
(methylnaphthalene, CH₃)	50
(phenanthrene-type structure)	55
(fluorene-type structure)	66
(fluoranthene-type structure)	80
(pyrene-type structure)	83
(dibenzofuran, O)	51
(dibenzothiophene, S)	66
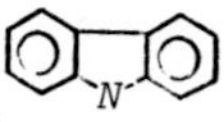 *(carbazole, N)*	87
Unheated Coal	20

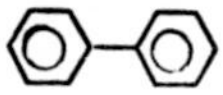

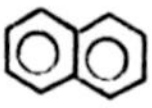

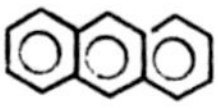

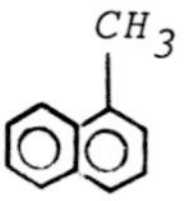

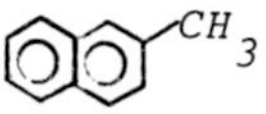

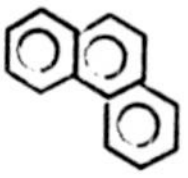

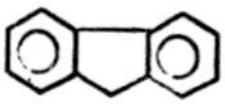

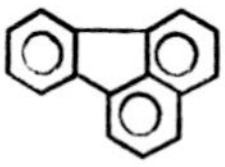

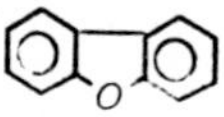

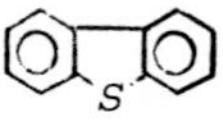

(Left margin vertical labels: HYDROCARBONS for rows 1–9; HETEROCYCLICS for the three heteroatom-containing rows.)

[a] Reference 9-25.
[b] Percent of coal which is soluble in quinoline after the reaction.

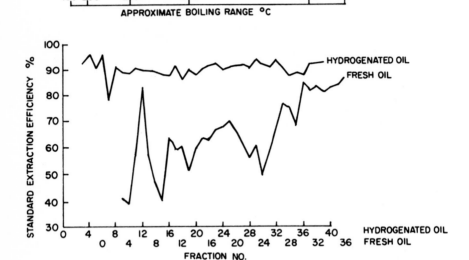

FIGURE 9-23. *Extraction Efficiencies of Fractions of Fresh and Hydrogenated Oil.*

synergistically when the content of H-donors is limited in the solvent.   In solvents of limited but measurable H-donor concentrations, correlations exist between the content of polyaromatic ring compounds and the extent of coal conversion at short times (9-2).   This is shown in Figure 9-24 for a series of coal-derived solvents with low H-donor capacities (.5 g H/100 g).   Little correlation was found between the extent of coal conversion and the content of H-donors alone in these solvents at short times.

Using these same solvents an interesting correlation was found between the conversion of coals at short times and the concentration of highly condensed aromatic rings, as measured by polarography.   In Figure 9-25 we show a plot of the percent coal conversion vs. the total current measured in the region of pyrene and fluoranthene reduction (9-34).   This agreement indicates the value of polarography for monitoring the quality of a solvent for short time conversion, where H-shuttling can be as effective as H-donation.

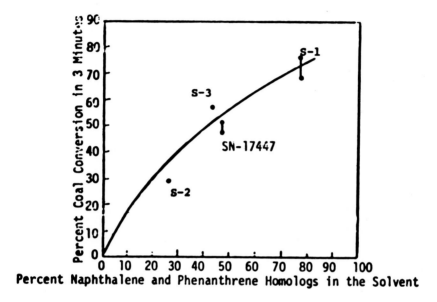

FIGURE 9-24. West Kentucky 9,14 Coal Conversion at 4 Minutes (740-780°F) in Solvents of Limited H-donor Capacity.

Not all aromatic molecules are good H-shuttlers, however. Pyridine is an excellent solvent for coal products, but cannot promote coal solubilization at either short or long times. This is probably because the molecule is not labile toward acceptance of "H" from the coal, or because hydrogen abstraction by coal radicals is not facile. Single ring aromatic compounds in general would have these limitations.

In Table 9-11, we present the data from a number of short residence time reactions between bituminous coals and aromatic solvents. Pyridine is less effective in promoting conversion to pyridine-soluble form than the more condensed aromatic solvents. Hydrogen gas was beneficial when pyridine was the solvent, but at

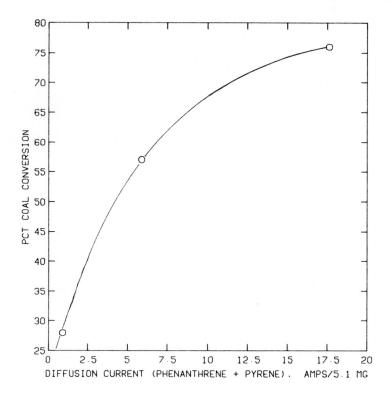

FIGURE 9-25.   Coal Conversion and the Polarographic Responses of Solvents.

the short times studied, did not promote higher conversions in methylnaphthalene or diphenyl ether solvent. In no case was a conversion of more than 53% observed with these slightly condensed or non-condensed ring systems.

C. The Effect of Phenols

Phenolic compounds constitute a major portion of the steady-state recycle solvent in SRC processes. The beneficial effect of phenols on coal liquefaction is well-documented in the literature (9-26,9-27,9-28). One caution should be noted in evaluating the effects of such components, however. Coal "conversion" is often used as the measure of effectiveness of a given solvent. The conversion, however, is often obtained by filtration of the crude product mix (sometimes hot) followed by washing of the

TABLE 9-11.   Conversion of Coals in Solvents
Without Hydrogen Donors

Solvent	Coal	H$_2$	Time (Min.)	Temp. (^{o}F)	% Conversion
Pyridine	WKY	-	1.3	800	31
Pyridine	WKY	-	60.0	796	30
Pyridine	WKY	X	1.2	800	40
Methylnaphthalene	BS	-	3.0	760	50
Methylnaphthalene	BS	X	3.0	775	49
Naphthalene					
Biphenyl	BS	-	3.0	763	53
p-Cresol	BS	-	2.3	796	53
Diphenylether	BS	X	3.3	800	53
Pyrene	MT	-	60.0	752	44
Quinoline	MT	-	60.0	752	40
Pyrene	PITTS	-	60.0	752	75

residue. If the wash solvent is a low polarity solvent, the
phenols will be observed to increase conversion only because they
have good physical solubility characteristics. Thus, if the
solvent contains some phenolic components, a significant portion
of the coal products will be made soluble for the filtration step
prior to treatment with a wash solvent.

Phenols greatly enhance the solubility of coal products in
the SRC process. As a result, even short-contact time products
(e.g., preheater conversions) can be filtered and coal conversions
of up to 90% have been observed in operating PDU's. Other processes such as the Exxon Donor Solvent Process (EDS) have solvents which contain much lower phenol contents as they are removed catalytically during the regeneration of the solvent by
external hydrogenation. As a result, the coal products do not
dissolve in the solvent until they are converted to a high-
quality product such as distillate.

Barring these physical changes in the solvent, we discuss
below how the presence of phenolic constituents in the recycle
solvent can affect the chemistry of the overall coal liquefaction
process.

The chemistry of this class of solvent components can be discussed in two parts: the chemistry of light phenols (≤9 carbons)
and the chemistry of heavy phenols (>9 carbons).

1. The Chemistry of Light Phenolic Components in Recycle
Solvents

The light phenols constitute the major portion of the phenolic
constituents of SRC recycle solvents and represent about 2/3 of
the total -OH content per unit weight of the solvent. These
lighter components do not contain hydroaromatics and would not be
considered as classical hydrogen donors.

In liquefaction reactions where the demand for hydrogen is
high and the solvent is limited in hydroaromatic content, the
light phenols can "donate" hydrogen through alkylation reactions
with aromatic rings. These alkylations can proceed through the
phenol nucleus, the -OH groups, or through alkyl side chains of
the phenols, and the phenols can become incorporated into coal
products. The chemical nature of such products will depend on
how the phenol is incorporated; if the phenolic group is retained
in an SRC product, for example, the SRC will be more polar, less
soluble, and will have increased sensitivity to charring (as discussed in Chapter 8). If the phenolic group is lost (e.g.,
through ether formation), the product will have a higher oxygen
content, but will not have reduced solubility.

Even when the hydroaromatic content of the solvent is relatively high, light phenols can become incorporated into coal
products and alter their chemical properties (particularly SRC).
To illustrate this, we present in Table 9-12, a comparison of the
reaction of Illinois #6 (Monterey) coal at short and long contact
times (800°F, 1000 psi H_2) with two synthetic recycle solvents.
Both solvents contained similar hydroaromatic concentrations
(40-50%). One solvent contained 18% p-cresol and the other had
none. The data show that in the initial phases of coal liquefaction, light phenols become incorporated into the SRC to give
a product of higher oxygen content. Some of the phenolic groups
are lost, but the end result is that the SRC is more highly
functional when light phenols are present in the solvent.

At longer reaction times, the oxygen content of the SRC and
the SRC functionality are nearly the same for both solvents.
This indicates some reversibility in the incorporation of phenols
with hydroxyl retention.

The amount of p-cresol lost from synthetic recycle solvents

TABLE   9-12.   *Comparison of SRC's From Phenol and Nonphenol-Containing Reactions*

(Illinois #6 Monterey Coal, 800°F, ~1000 psi H_2)

					Increases due to Phenol	
Contact Time (min)	4	4	90	90	at 4	at 90
-OH (meq/g)	3.4	2.8	2.1	2.2	0.6	-
Oxygen (%)	9.6	8.9	5.8	5.4	0.7[a]	0.4
SRC Yield (%)	82.84	77.92	68.60	64.00	4.92[b]	4.60
Balance (%)	101.60	96.69	98.85	97.49	4.91	1.36
		Solvent Composition (%)				
Tetralin	41	50	41	50		
p-Cresol	17	0	17	0		
Methylnaphthalene	40	49	39	49		
γ-Picoline	1.1	0	1.3	0		

[a] *Corresponds to an increase of 0.43 meq O/g solvent.*

[b] *Corresponds to an increase of 0.46 meq O/g solvent.*

was rather small when the solvent also contained high concentrations of hydroaromatics. However, as discussed earlier, when the hydroaromatic H-donor content was limited, coal radicals abstracted hydrogen from other sources. When H-abstraction occurred from p-cresol, the resultant radical became stabilized through several routes, a number of which led to the conversion of p-cresol to other species. Figure 9-26 shows the extent of conversion of p-cresol as a function of time of reaction for solvents of varying tetralin contents.

The major conversion products of p-cresol were phenol and alkyl phenols. The amount of these products increased with decreasing tetralin content of the solvent (800°F, 90 minutes, 1000 psi H_2) and paralleled the production of 2-methylnaphthalene dimers discussed earlier. In most cases the yields of phenol and alkyl phenols were equal. The alkyl phenols were found by GC-mass spectroscopy to be predominantly ethyl phenol.

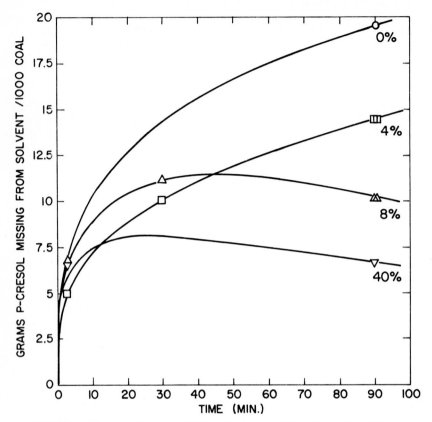

FIGURE 9-26.   p-Cresol Conversion vs. Time in a Wyodak Coal Conversion.

A speculative mechanism for the production of these two conversion products has been proposed and is given in Figure 9-27. Alternatively, the cresol benzylic radical could add to another p-cresol aromatic ring and then cleave to produce 2,4-dimethylphenol and phenol. By a similar mechanism p-cresol radicals can react and condense with coal or coal products to yield SRC's with different properties than non-condensed SRC's.

The conversion of p-cresol to disproportionated products or coal condensation products is not necessarily an undesirable reaction. In hydroaromatic-deficient solvents, p-cresol can provide hydrogen to the reactive coal radicals. The result is that higher coal conversions (to pyridine-soluble products) are observed in solvents containing p-cresol than in solvents which have none (see Table 9-12).

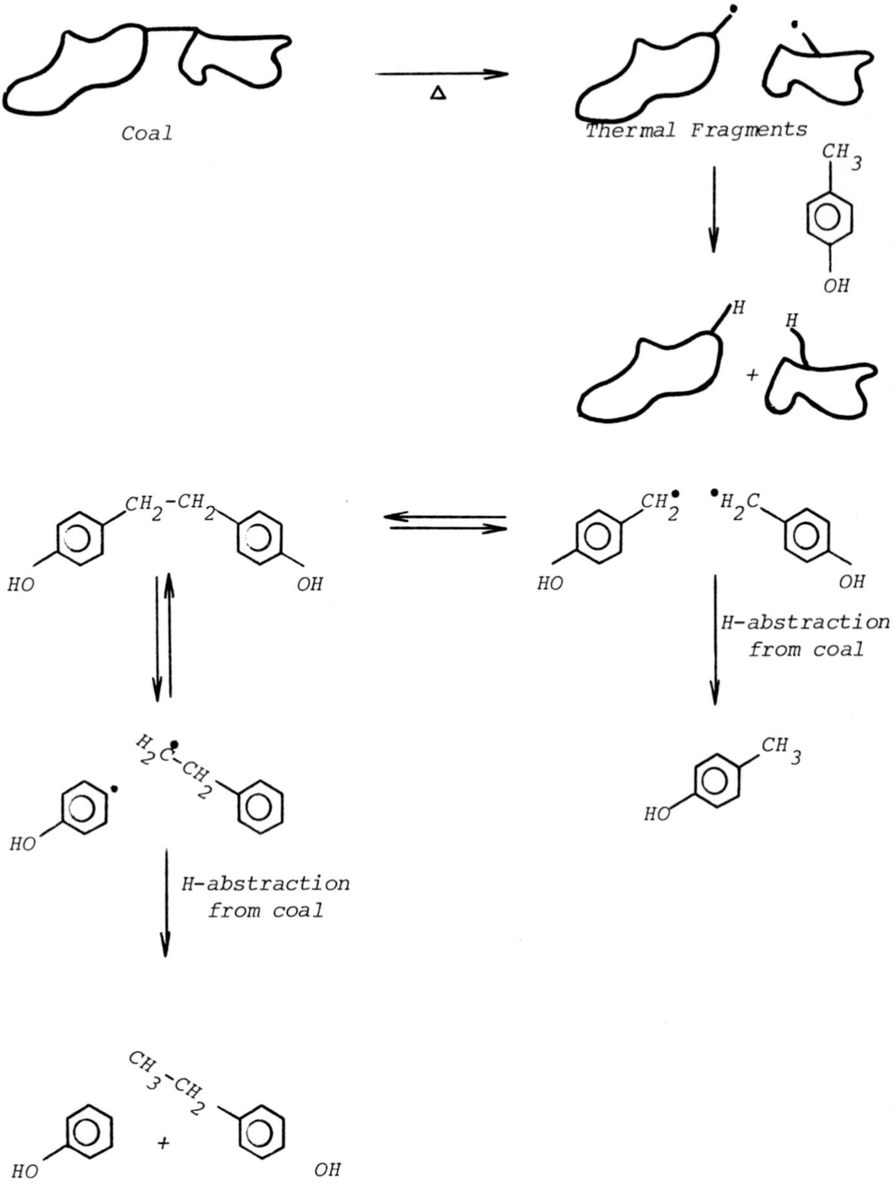

FIGURE   9-27.   Speculative mechanism for the conversion of
                 p-cresol to phenol and alkyl phenols.

2. The Chemistry of Heavy Phenolic Components in Recycle
Solvents

The chemistry of the heavy phenolic components of recycle solvents is less well understood. Clearly, the mechanistic routes
available to light phenols are also available to heavy phenols;
however, two additional interactions are also possible. Heavy
phenols having more than 9 carbons have the potential of hydroaromatic structure and could thus function as hydrogen donors via
dehydrogenation. The more functional heavy phenols can also form
chars.

Some work has been devoted to characterization of the heavy
phenols of typical recycle solvents to aid in understanding their
chemistry (9-34).

From elemental analyses shown below, the heavy phenols are
rich in hydrogen and relatively low in aromatic carbon content.

TABLE 9-13. *Composition of Heavy Phenols*

Elemental Analysis (Wt%)				% Aromatic		Molecular Weight		
C	H	O	N	C	H	VPO[a]	FIMS[a]	FIMS[b]
81.26	7.6	9.43	1.2	72	40	175	244	320

[a] *Number average molecular weight.*

[b] *Weight average molecular weight.*

Field Ionization Mass Spectroscopic analysis of these materials
indicated that high concentrations of hydroxytetralin homologs are
indeed present (9-34). These heavy phenols are likely to have the
potential to act as classical hydrogen donors.

Hydroaromatic phenols have been found to be excellent H-donors
in coal conversion (9-34). Illinois #6 (Burning Star) coal was converted with and without hydroxytetralin as a component of synthetic
recycle solvents. One solvent was composed of 70% methylnaphthalene, 15% tetralin and 15% 6-hydroxytetralin. This mixture contained 1.1 meq/gm phenolic -OH and .9 g hydroaromatic H/100 g solvent. The other solvent contained 43% tetralin and 57% methylnaphthalene.

Figure 9-28 shows the results of conversions at 800°F, 1200
psi H_2 with samples withdrawn up to 90 minutes. It can be seen
that 6-hydroxytetralin was 3 times more active for donation of
"H" to coal than tetralin.

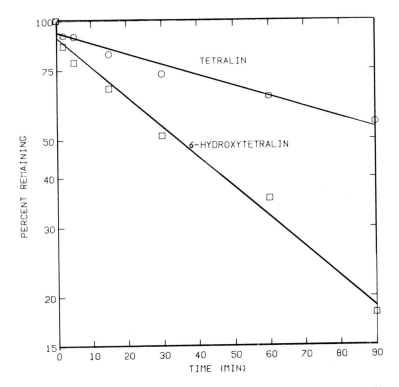

FIGURE 9-28.   Conversion of 6-Hydroxytetralin and Tetralin

This reaction had an additional complication, however. The product β-naphthol was not stable in the reaction medium and was further dehydroxylated to form naphthalene.

This dehydroxylation consumed H with the result that 6-hydroxytetralin on extended reaction is a net donor of only two hydrogens per molecule rather than four. Dehydroxylation was observed only when coal was present and is thus catalyzed by either the coal or intrinsic coal minerals. The rate of β-naphthol dehydroxylation was about equal to the rate of tetralin dehydrogenation in this study.

When the above conversion is compared with a conversion in which tetralin was the only hydroaromatic and p-cresol was the

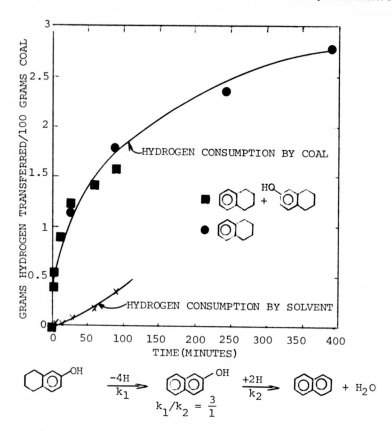

FIGURE 9-29.   Hydrogen Consumption From Solvent.

phenol, the rate of hydrogen consumption by the coal was essentially identical (see Figure 9-29). This observation points out
that the coal, and not the hydrogen donor, is responsible for the
rate of hydrogen donation. In addition to the coal's consumption
of hydrogen, the dehydroxylation reaction also required hydrogen.

   Other hydroxy positional isomers have similar chemistry.
5-Hydroxytetralin is also about 3 times more active as a hydrogen
donor than tetralin, but the rate of α-naphthol dehydroxylation
is about 15 times faster than the rate of tetralin dehydrogenation. If the hydroxy group is on the saturated ring, the molecule very rapidly disproportionates to water, naphthalene, and
tetralin.

The loss of hydroxy groups from phenols in the solvent can also occur through the formation of ethers or other condensation reactions. α-Naphthol is known to convert to a furan derivative in up to 80% yields through thermal reactions at temperatures similar to those of liquefaction (9-29).

We have observed as much as 18% of a dimeric species formed from phenols in some cases, but the structures have not been determined (except for p-cresol; see earlier in this section).

Reactions such as these may or may not contribute net additional hydrogen, but they definitely increase the molecular weight of the products and may also be significant to the formation of char from highly phenolic SRCs or solvents.

D.   The Effect of H-Acceptors

The beneficial effects of H-donors in recycle solvents were discussed above. High concentrations of H-donors generally produce higher quality SRC's (higher H, lower O, S, and aromatic C). If, however, some solvent components can consume hydrogen, then the overall hydrogen available for reaction with coal is reduced. Reactions which lead to the demand for hydrogen by solvent components include cracking to form light hydrocarbon gases, reduction in molecular weight of heavy components and dehydroxylation. Since coal or SRC can function as H-donors, any H-acceptor in the solvent can lower the quality of the SRC. As discussed in Chapter 8, H-acceptors promote char formation from SRC components or even solvent components, and lower apparent conversions are observed.

Table 9-14 shows that a synthetic solvent with 40% tetralin when admixed with an equal weight of benzophenone (an H-acceptor) did not form char. However, a process-derived solvent is quite sensitive to the presence of H-acceptors under liquefaction conditions and can produce as much as 21% char in 2.6 hours. If the H-donor level is raised slightly by the addition of 3% tetralin, the charring tendency of this solvent is considerably lowered. The same trend has also been reported for whole SRC or fractions therefrom.

TABLE   9-14.   *Acceleration of Charring by H-Acceptors*

Solvent	Pure Component	With 50% Benzophenone	With 47% Benzophenone 3% Tetralin
(% Char After 2.6 Hours at 850°F)			
SS*40	0	0	–
SN-6666	Negligible	21	12

Monterey SC-SRC	Pure Component	With 14-17% Benzophenone	With 14% Tetralin
(% Char After One Hour at 850°F)			
Whole	43	51	2
Benzene Insoluble	76	89	46

The effect of H-acceptors on coal liquefaction was tested by reacting coal at 800°F for 3 hours with a solvent which had no hydroaromatics but contained 16% benzophenone and 17% p-cresol. A very low "conversion" was observed and the balance was 150%. The recovered solid product was shown to be essentially mesophase by optical microscopic analysis (9-31). The above results indicate that when solvents contain H-acceptors, the solvents themselves are unstable and that char formation from SRC components or coals is promoted.

E.   Char Formation From Solvent

One problem in coal liquefaction which plagues plant operators is the formation of solid products from soluble species. The organically derived solids have been referred to as char, coke, mesophase carbon, etc. Such solids are commonly accounted for as unconverted coal as they are physically inseparable from it. A detailed discussion of the chemistry of char formation was given in Chapter 8; however, we will summarize here a few points which are pertinent to the chemistry of the recycle solvent.

Generally, char production is not desired as "conversions" are lowered and plant operations can be impaired by line blockage or the build up of solids in the reactor. In principle, char formation is not all bad. For example, the conversion of some SRC (∿6% H) to char (∿4% H) can supply hydrogen when solvents are limited in hydroaromatics content. Filtration can also be improved if small particles grow to larger ones (by char deposition). In any event liquefaction processes will need some highly carbonaceous solids for the production of hydrogen gas via gasification (high surface areas are necessary, however).

Thus, what appears to be important in liquefaction processes is not whether or not char forms but rather how it can be controlled.

The tendency toward char formation of SRC components appears to be closely related to high chemical functionality. We have previously reported (9-1) that the predominant initial products of coal dissolution are benzene insoluble, high in molecular weight, and high in phenol content. These materials are much more prone to char formation than the less functional benzene or hexane soluble products which predominate at longer reaction times (see Table 9-14).

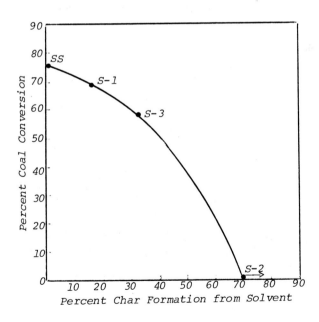

FIGURE 9-30. Solvent Char Formation vs. Coal Conversion

Solvents can also be converted to char, leading to a lower observed "conversion" of coal and high balances based on coal fed. We have reported (9-2) that there exists an inverse correlation between the conversion of a coal under standard conditions and the char yield of solvents on programmed pyrolysis (via TGA). This is shown in Figure 9-30.

The process conditions can also influence char formation. Char yields increase with higher temperatures, longer contact times, or altered solvent compositions with lower H-donor content or higher H-acceptor or phenol levels.

F. Coal Conversion With Process-Derived Solvents

Most of the coal conversions discussed above used synthetic recycle solvents. We discuss here a few examples of conversions in process-derived solvents for comparison.

In typical SRC recycle solvents, characterization work showed (9-2) that the hydroaromatic hydrocarbons were predominantly hydronaphthalene analogs. Hydrophenanthrenes were much lower in concentration. The high reactivity of such compounds would, of course, limit their content at steady state. The total available "H" from hydroaromatic hydrocarbons was estimated to be .1-.3 g H/100 g solvent. In addition, the heavy phenols can potentially provide another .1-.3 g H/100 g solvent as discussed above. Coal conversions with heavy phenol and heavy non-phenol fractions and some full-range solvents are summarized and compared to conversions with synthetic solvents in Table 9-15. The whole recycle

TABLE 9-15. *Conversions with Processed-Derived Solvents*

Monterey 90 Minutes

Solvent	SS*40	SS*8	Whole Ky 650°F⁻	Heavy Non-phenols 450–850°F	Heavy Phenols 450–850°F
Conversion	92.0	84.5	82.4	87.1	43.3

Kentucky 40 Minutes

Solvent	SS*40	Full Range Burning Star
Conversion	92.5	88.0

solvents gave conversions similar to a limited H-donor solvent
(e.g., about 8% tetralin). The heavy non-phenols were somewhat
more effective than the full-range solvent. Heavy phenols, on
the other hand, gave low apparent coal "conversions".

This low apparent conversion is believed due to solvent charring, resulting from H-donation from the heavy phenols. These
materials did not char in the absence of coal (as determined by
blank reactions), but when the heavy phenols were admixed with
benzophenone and reacted under the same conditions as the above
coal conversions, large quantities of char were produced. It
should be noted that the SRC yields were very similar for conversions using heavy phenol and heavy non-phenol solvents. The
balance, however, was very high for the heavy phenol solvent. In
the order of 18% of the solvent being converted to char would
explain the high balance.

Balances for whole recycle solvents were good, so it appears
that heavy phenols do not char extensively when other hydrogen
donors are also present in the solvent. It may be that many of
the reported instances of poor performance observed in pilot
operations can be explained by solvent component charring due to
limitations in the content of hydroaromatic hydrocarbons.

A possible example of this is shown in Table 9-16 where two
solvents were obtained from the Wilsonville PDU during "good" and
"poor" performance for the conversion of Illinois #6 (Burning
Star) coal. The estimated available "H" from the whole solvents
was similar but the "poor" solvent contained more heavy phenols
and less hydroaromatic hydrocarbons.

Compositional changes which occur in true recycle solvents on
reaction with coal were also investigated. (Illinois #6 (Monterey),
$800°F$, 1000 psi H_2, solvent/coal = 4.9). A West Kentucky 9,14

TABLE 9-16 . *Relationships Between H-Donor Content and
Illinois #6 (Burning Star) Coal Conversions at Wilsonville*

Solvent	Estimated Available "H" Content (g H/100 g solvent)			Reported Conversion %	Comments
	Hydroaromatic Hydrocarbons	Heavy Phenols	Total		
6663	.24	.27	.51	95	Stable performance
6540	.13	.40	.53	75-85	Erratic performance

recycle solvent was characterized by our liquid chromatographic techniques (9-1, 9-2) before and after the reaction. These analyses showed that the major change which occurred was the transformation of monoaromatic ring hydrocarbons (e.g., tetralins) into diaromatic ring hydrocarbons (see Table 9-15). Almost no change was observed in the content of saturates, tri- and higher aromatic hydrocarbons, or phenols. Gas chromatographic analyses of the phenols also showed no major changes.

G.   Regeneration of the H-Donors in Recycle Solvents

In Chapter 6 we discussed the results of our studies on the catalytic effects of mineral matter on reactions pertinent to solvent hydrogenation. In the discussions below we summarize some of our observations on the factors critical to such hydro-genations. The solvent on reaction with coal changes in compo-sition (primarily hydroaromatics are converted to condensed aromatics). If a stable steady state operation is to be achieved, these changes must be reversed to restore the original solvent composition. Two means of restoring this composition are avail-able. Coal can be converted to provide new additional solvent or the hydrogen-depleted solvent can be rehydrogenated by gaseous hydrogen. In the SRC process both means are operative but both are relatively slow when compared to coal conversion to soluble form. Figure 9-31 shows that solvent range products of coal

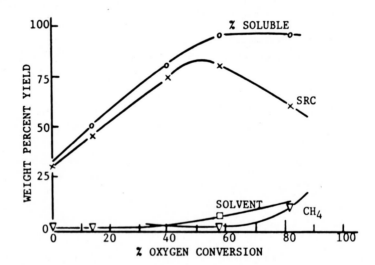

FIGURE 9-31.   Product Yields vs. Percent Oxygen Conversion for West Kentucky Coal.

become appreciable only after about 50% of the oxygen has been removed from a West Kentucky 9,14 bituminous coal. At the same time, light hydrocarbon production also becomes significant.

The rehydrogenation of solvents by hydrogen gas is potentially subject to a number of controlling factors, but thermodynamics does not appear to be a limitation. Our calculations show that naphthalene $\rightleftharpoons$ tetralin and phenanthrene $\rightleftharpoons$ hydrophenathrene equilibria are quite favorable. Yet in all of the SRC solvents analyzed, the hydroaromatic concentrations are well below that allowed by thermodynamics; thus the problem is one of rate. Solvent rehydrogenation is slow relative to solvent H-donation to coal or coal products. This point exphasizes the importance of maintaining a high level of hydroaromatic hydrocarbons in the solvent, as on reaction with coal the demand for "H" will be the same independent of the availability of good H-donors. As shown above, if the hydroaromatics are limited, side reactions such as phenol condensation, solvent dimerization, or char formation will occur. This is illustrated in Figure 9-32.

If solvents are to be regenerated through interaction with hydrogen gas, it is generally agreed that catalysts must be involved. But just what are these catalysts in SRC processes?

The intrinsic mineral matter in coal can provide a source of metals; pyrites are commonly believed to be the most important (9-32). Another suspected catalyst is reactor solids which accumulate in the dissolver. Large heavy solid particles will not be carried out; thus, there is a natural accumulation of high ash solids (usually agglomerates).

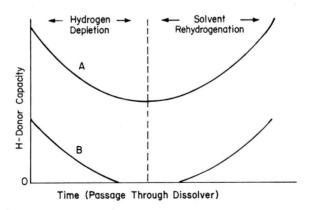

FIGURE 9-32. Dynamics of Solvent Hydrogen Donor Capacity During Coal Liquefaction.

Attempts to measure the catalytic activity of high iron reactor solids produced at the Wilsonville PDU were described in
Chapter 6. Synthetic solvent mixtures were heated under typical
liquefaction conditions with 5 weight percent solids and the results compared with those of blank runs and $Co/Mo-Al_2O_3$ catalyzed
runs. The rates of equilibration of tetralin $\rightleftharpoons$ naphthalene and
methylnaphthalene $\rightleftharpoons$ methyltetralin over these catalysts were
found to be indistinguishable from the thermal blank run in these
experiments; thus, the reactor solids level of activity for
hydrogenation-dehydrogenation was rather low. In a PDU-scale SRC
process, the temperature and pressure are somewhat higher, the
solids are kept in sulfided form, and the amount in the reactor
higher. Under these more severe conditions, it is possible that
some catalytic rehydrogenation could be observed but it is felt
that the major catalytic species for solvent rehydrogenation may
be something other than reactor solids.

It should be recalled that dehydroxylation of polyaromatic
phenols is rather easy to catalyze and reactor solids most
probably play a role in this reaction. In addition, evidence was
found that dehydroxylation of monoaromatic phenols is catalyzed
by reactor solids. Reactions such as this can account for the
conversion of SRC to solvent range products, increased hydrogen
consumption, and improvement in the SRC quality.

The mineral matter in coal may also be important in
hydrogenation-dehydrogenation reactions in that it is in a finely
divided state and could have appreciable surface area. In one
experiment coal was reacted with a synthetic solvent containing
tetralin and 9,10-dihydrophenanthrene. The 9,10-dihydrophenan-
threne reacted very rapidly; however, on extended reaction the
concentration of 9,10-dihydrophenanthrene appeared to approach a
steady state value of about 7% of the phenanthrene products (see
Figure 9-7). In addition, the concentration of 1,2,3,4-
tetrahydrophenanthrene continually increased in this run to a
value of 3.7% of the phenanthrene products by the time the reaction was terminated. These observations do indicate that some
hydrogenation catalysis was operating in this run and coal minerals may have been responsible.

A more dramatic example of the effect of catalysis by intrinsic mineral matter in coal was observed when conversions were
conducted with demineralized and unaltered Illinois #6 (Burning
Star) coal. Demineralization consisted of low temperature extractions of the coal with dilute HF/HCl followed by pyrite removal by the courtesy of TRW using the Meyers process (9-33). The
comparison conversions were conducted in a synthetic solvent
which contained no hydroaromatics; thus consumption of hydrogen
from the gas phase was critical. In Table 9-17 we show that for
the untreated coal, removal of the hydroaromatics from the solvent

TABLE 9-17 . Effect of Mineral Matter on Conversion
in Non-Donor Solvents
Illinois #6 (Burning Star) Coal, 800°F, 90 minutes, ~1000 psi H_2

Percent Tetralin in solvent	43	0	0
Coal Treatment	None	None	Demineralized
Percent Conversion	89	69	40

lowered the conversion in a manner similar to that of the
Illinois #6 (Monterey) coal studies described earlier. When the
coal was treated to remove any potentially catalytically active
mineral matter, the observed conversion was considerably lower.
    The above results indicate that catalytic reactions are in-
deed important in SRC processes and that both the finely divided
mineral matter and the massive reactor solids are active. The
types of reactions catalyzed by these two materials may be
different, however.

H. Conclusions

    In SRC processes, the chemical nature of the recycle solvent
controls product selectivity in addition to establishing rates of
reaction. The rate of coal dissolution is responsive to the concentration of hydroaromatic hydrocarbons in the solvent as thermally generated radicals must be capped by hydrogen to produce
soluble species. The nature of the coal itself (energy distribution of bonds, functionality content, and skeletal structure)
determines the rate at which conversions can take place and the
hydrogen consumption that is required. The solvent can act to
provide hydrogen in various ways. If hydroaromatics are labile
and plentiful, they provide the majority of hydrogen that the
coal demands. High concentrations of active hydroaromatics in
the solvent also lead to more rapid desulfurization and deoxygenation.
    If hydroaromatics are limited, phenols and aromatic methyl
groups provide hydrogen through alkylation reactions. Light
phenols are predominantly added to coal products and heavy phenols
form some char. Aromatic methyl groups or other alkyls can dimerize forming ethylene linkages or can condense to form methylene
bridges or more condensed rings. In addition, hydrogen gas plays
an important role as thermally produced radicals can interact

directly with molecular hydrogen at liquefaction temperatures.
(9-5(a)).

Low hydroaromatic content solvents do not achieve as high
conversions as high hydroaromatic content solvents. If the same
conversions are achieved through more severe reaction conditions,
high yields of light hydrocarbon gases result. The SRC itself
may even be called on to donate hydrogen to the pool. When the
SRC must give up hydrogen, the product is of course lower in
quality and a portion is transformed into char.

Hydrogen donation by SRC, coal or residue can be aided by the
action of H-shuttlers which can transfer hydrogen from one portion
of the coal to another. The most efficient H-shuttlers appear to
be higher molecular weight, more highly condensed, aromatic hydrocarbons.

Recycle solvents can contain H-acceptors as well as H-donors;
the result of which is the same as a lower H-donor capacity (e.g.,
char formation, low conversion, etc.). Char formation per se is
not necessarily bad as long as it can be controlled. The reaction
can be a source of hydrogen and small particulates can be grown
larger to aid in filtration.

High char rates are associated with high temperatures, long
times, low hydroaromatic solvents, high functionality (hydroxyl
content) and high aromatic content in SRC or solvent species.

For a stable continuous operation, the hydrogen donors in a
recycle solvent must be capable of regeneration under coal conversion conditions. Aromatic $\rightleftharpoons$ hydroaromatic equilibria are
the predominant sources of reversible H-donation. Regeneration
of such H-donors appears to be kinetically, not thermodynamically,
limited. Some catalysis by intrinsic mineral matter or reactor
solids does aid in this regeneration but activities are low.

REFERENCES

9- 1.   D. D. Whitehurst, M. Farcasiu, and T. O. Mitchell, "The
        Nature and Origin of Asphaltenes in Processed Coals,"
        EPRI Report AF-252, First Annual Report Under Project
        RP-410, February 1976.

9- 2.   D. D. Whitehurst, M. Farcasiu, T. O. Mitchell, and J. J.
        Dickert, Jr., "The Nature and Origin of Asphaltenes in
        Processed Coals", EPRI Report AF-480, Second Annual Report
        Under Project RP-410-1, July 1977.

9- 3.   (a) R. Raymond, I. Wender, and L. Reggel, Science, 137,
        681-2 (1962).
        (b) L. Reggel, C. Zahn, I. Wender, and R. Raymond, U. S.
        Bureau of Mines Bulletin, 615 (1965).

9- 4.   (a) G. P. Curran, R. T. Struck, and E. Gorin, I&EC Process
        Design and Development, 6, 166 (1967).
        (b) E. Gorin, H. E. Lebowitz, C. H. Rice, and R. T. Struck,
        Proceedings, Eighth World Petroleum Congress, 4, 43 (1971).

9- 5.   (a) L. E. Furlong, E. Effron, L. W. Vernon, and E. L.
        Wilson, paper presented at National AIChE Meeting, Los
        Angeles, California, November 18, 1975.
        (b) L. E. Furlong, E. Effron, L. W. Vernon, and E. L.
        Wilson, Chem. Eng. Prog., 72, 71 (1976).
        (c) W. R. Epperly, "EDS Coal Liquefaction Process Develop-
        ment, Phases IIIB/IV", Annual Technical Progress Report
        for July 1, 1977-June 30, 1978. FE-2893-1T.
        (d) P. S. Maa, U. S. Patent 4,022,680, May 10, 1977.
        (e) R. C. Green and R. L. Dubell, U. S. Patent 4,084,054,
        September 13, 1977.
        (f) K. W. Plumlee and L. W. Vernon, U. S. Patent 4,051,012,
        September 27, 1977.
        (g) Exxon Research and Engineering Co., "EDS Coal Lique-
        faction Process Development" Phase IIIA, Final Technical
        Report for Period January 1, 1976-June 30, 1977. Vol. 1,
        p. 153 (calculations of meq./gm by D. D. Whitehurst).

9- 6.   L. Petrakis and D. W. Grandy, Preprints Fuel Division,
        146th National ACS Meeting, Miami, Florida, November 1978,
        p. 147.

9- 7.   J. W. Prather, A. M. Ahangar, W. D. Pitts, J. P. Henley,
        A. R. Tarrer, and J. A. Guin, Ind. Eng. Chem. Process
        Design and Development, 16, 267 (1977).

9- 8.   J. Uebersfeld, A. Etienne, and J. Combrisson, Nature, 174,
        614 (1954).

9- 9.   D. J. E. Ingram, J. G. Tapley, R. Jackson, R. L. Bond, and
        A. R. Murnaghan, Nature, 174, 797 (1954).

9-10.   H. L. Retcofsky, J. M. Stark, and R. A. Friedel, Anal.
        Chem., 40, 1699 (1968).
9-11.   (a) D. E. G. Austen, PhD Thesis, University of Southampton,
        1958.
        (b) D. E. G. Austen and D. J. E. Ingram, Conf. on Sci. Use
        Coal, Sheffield 1958, A-39.
        (c) D. E. G. Austen, D. J. E. Ingram, P. H. Given, and M.
        E. Peover, Fuel, 38, 309 (1959).
        (d) D. E. G. Austen, D. J. E. Ingram, P. H. Given, C. R.
        Binder, and L. W. Hill, "Coal Science", Advances in
        Chemistry, Series #55, ACS (1965) p. 344.
9-12.   J. Duchesne, J. Depireux, and J. M. van der Kaa, Geochim.
        Cosmochim. Acta, 23, 209 (1961).
9-13.   D. J. E. Ingram, "Free Radicals as Studied by Electron
        Spin Resonance", Butterworth, London, 1958.
9-14.   D. W. van Krevelen, "Coal", pp. 394-399, Elsevier,
        Amsterdam, 1961.
9-15.   C. Kroger, Brennstoff-Chem., 39, Sonderausgabe, 62-7 (1958).
9-16.   E. de Ruiter and H. Tschamler, Brennstoff-Chem., 42, 290
        (1960).
9-17.   L. S. Singer, "Proceedings of Fifth Carbon Conference",
        Vol. 2, p. 37, Pergamon Press, Oxford, 1963.
9-18.   J. Smidt and D. W. van Krevelen, Fuel, 38, 355 (1959);
        Coal, pp. 393-399, Elsevier, Amsterdam (1961).
9-19.   B. C. Gerstein, C. Chow, R. G. Pembleton, and R. C. Wilson,
        J. Phys. Chem., 81, 565 (1977).
9-20.   (a) L. Petrakis and D. W. Grandy, Anal. Chem., 50, 303
        (1978).
        (b) L. Petrakis and D. W. Grandy, Preprints Fuel Division,
        146th National ACS Meeting, Miami, Florida, November 1978,
        p. 147.
9-21.   D. Kang, L. L. Anderson, W. H. Wiser, Preprints Fuel
        Division, 173rd National ACS Meeting, New Orleans,
        Louisiana, March 1977, p. 160.
9-22.   (a) R. C. Neavel, Proc. Symp. on Agglomeration and Con-
        version of Coal, Morgantown, West Virginia, 1975.
        (b) R. C. Neavel, Fuel, 55, 237 (1976).
9-23   (a) C. Golumbic, J. E. Anderson, M. Orchin, and H. H.
        Storch, U. S. Bureau of Mines Report Investigation 4662
        (1950).
        (b) M. Orchin, C. Golumbic, J. E. Anderson, and H. H.
        Storch, U. S. Bureau of Mines, Bulletin 505 (1951).
9-24.   (a) L. A. Heredy and P. Fugassi, "Phenanthrene Extraction
        of Bituminous Coals", in ACS Monograph #55, "Coal Science",
        p. 488.

(b)   J. J. Ratto, L. A. Heredy and R. P. Skowronski,
Division of Fuel Chemistry Preprints, CJS/ACS Chemical
Congress 1979, Honolulu, Hawaii.

9-25.   G. O. Davis, F. J. Derbyshire, and R. Price, J. Inst.
Fuel, 50, 121 (1977).

9-26.   A. Pott and H. Broche, Fuel, 13, 125 (1934).

9-27.   M. Orchin and H. H. Storch, Ind. Eng. Chem. 40, 3225 (1975).

9-28.   Y. Kamiya, H. Sato and T. Yao, Fuel, 57, 681 (1978).

9-29.   C. Collins, Oak Ridge National Laboratory, Oak Ridge,
Tenn., personal communication.

9-30.   P. L. Walker, Jr., W. Spackman, P. H. Given, A. Davis,
R. G. Jenkins and P. C. Painter, "Characterization of
Mineral Matter in Coals and Coal Liquefaction Residues",
EPRI Project 366-1, AF-832, December 1978.

9-31.   G. Mitchell, personal communication, The Pennsylvania
State University, Department of Mineral Sciences.

9-32.   T. O. Mitchell, From Discussions at the Meeting on
"Mineral Matter Effects and Use of Disposable Catalysts
in Coal Liquefaction", Sandia Laboratories, Albuquerque,
New Mexico, June 14-15, 1978.

9-33.   R. A. Meyers, "Coal Desulfurization", Marcel Dekker, Inc.,
New York, 1977.

9-34.   D. D. Whitehurst, T. O. Mitchell, M. Farcasiu, and J. J.
Dickert, Jr., "The,Nature and Origin of Asphaltenes in
Processed Coals", EPRI Research Project 410-1,
Third Annual Report and Overall Summary, March 1977 -
January 1979.

9-35.   M. Farcasiu, Fuel, 56(1), 9 (1976).

9-36.   J. G. Bendoraitis, A. V. Cabal, R. B. Callen, T. R. Stein,
and S. E. Voltz, "Upgrading of Coal Liquids for Use as
Power Generation Fuels", First Annual Report to EPRI Under
Project RP-361, January 1976.

9-37.   E. Leete, JACS, 83, 3645 (1961).

9-38.   S. Yokoyma, N. Ounisi and G. Takeya, Nikka, 10, 1963 (1973).

9-39.   J. P. Teas, J. Paint Technology, 40, 19 (1968).

9-40.   A. D. Sherry and K. F. Purcell, J. Phys. Chem., 74, 3535
(1970).

9-41.   B. L. Karger, L. R. Snyder, and C. Horvath, An Introduction
to Separation Science, Wiley, New York, 1973.

9-42.   J. H. Hildebrand and R. L. Scott, "The Solubility of Non-
Electrolytes", Third Edition, Reinhold Publishing Company,
New York, 1949.

9-43.   C. M. Hanson, J. Paint Technology, 39, 39 (1967); Ind. Eng.
Chem., Prod. Res. and Develop., 8, 2 (1969).

9-44.   I. Mochida, A. Takarabe, K. Takeshita, Fuel, 88, 17 (1979).

Chapter 10

LIMITATIONS OF CURRENT LIQUEFACTION TECHNOLOGY

AND THE SIGNIFICANCE OF RECENT DEVELOPMENTS

I. INTRODUCTION

This chapter discusses how presently developing coal liquefaction processes are limited by intrinsic chemical constraints
which are imposed by temperature, catalyst, solvent quality and
the chemical structure and functionality of coals being processed.
In addition, the significance of recent research to developing new
or improved coal liquefaction technologies will be covered.

II. HISTORY OF DEVELOPING PROCESSES

Coal was the major fossil fuel until the early twentieth
century when it was substantially replaced by petroleum and
natural gas. Now as a result of diminishing and unreliable
supplies, as well as increased demand and costs of these materials,
coal may once again become our most important fossil fuel source.
However, our distribution and process systems are now substantially based upon liquids and gases. For environmental reasons,
all fuels must be low in sulfur. Consequently, there is a need
for the technology to convert coal both to low ash and sulfur
solids and to liquids and gases.

Berthelot (10-1) first reported in 1869 that coals could be
converted to liquids by hydrogenation. Almost 50 years later
Bergius (10-2) patented a high-pressure, high-temperature hydro-
extraction process. During World War II, Germany successfully
converted up to 9000 tons/day of coal in catalytic processes
involving long-contact time, high hydrogen pressure (600-700 atm)
and high temperature (up to 890°F).

There has also been significant research in coal liquefaction in Japan. Past research centered on electrode production for their aluminum industry. Since this required ash removal only, a non-hydrogenative dissolution process was developed using a highly aromatic, essentially non-donor, solvent without H_2 gas (10-3). Yields of SRC of up to 80% were reported from high-volatile bituminous coal, though in the commercial plant the yield was only 55 percent. This SRC was then carbonized to produce electrode coke. One 250-ton per day plant was built, but World War II stopped the program; interest is now reviving. The process is unusual in that highly hydrogenated products are not desired and only mineral matter removal is important. Thus, hydrogen manufacture is not a critical feature.

The German technology, while not economically competitive in today's market, has been the basis for most of the recent research efforts in the United States. The emphasis is on conducting essentially the same reactions under less severe conditions. Ultimately all modifications are directed toward improving thermal efficiency and hydrogen utilization, reducing capital investments and operating costs, broadening the range of acceptable feed-coal properties, and increasing the product slate flexibility.

It is certain that none of the processes presently in an advanced stage of development is ideal; all can be improved upon. It is possible, however, that one or more, with modifications, will be used commercially. It is also probable that the process that ultimately will be the workhorse of this industry has not yet been developed. A great deal of research is still very much needed.

The number of processes and process variations designed to convert coal (or other similar solid fuels such as peat, lignite, etc.) to cleaner, more conveniently useful fuels has been increasing rapidly in recent years. However, space considerations make it impossible to give more than a brief description of each of the potentially useful processes. We will limit our concern only to direct liquefaction processes which are or have been considered for commercial development in the United States:

1. SRC-I and SRC-II (PAMCO and GULF)

2. Exxon Donor Solvent Process

3. CFS Two-Stage Liquefaction (CONOCO)

4. H-Coal (HRI)

5. Zinc Halide Coal Liquefaction (CONOCO-SHELL)

6. Dow Coal Liquefaction Process

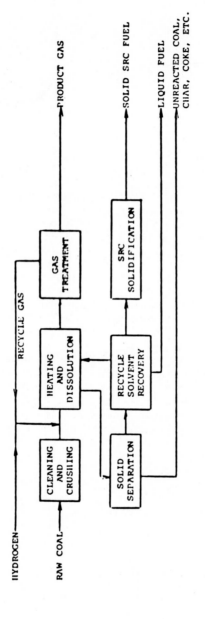

Process Description

The SRC-I Process is in essence a refinement of the POTT-BROCHE Process. No Catalyst. Ground coal is mixed with recycle solvent and hydrogen. Slurry is preheated, then reacted 30-90 minutes at 800°F, 1000-2000 psig operating pressure. Char, mineral matter and unreacted coal are removed by filtration or solvent precipitation. Solvent is recovered by distillation. A solid fuel which can be melted at 350°F and is low in sulfur and ash content is the primary product.

Figure 10-1(a). Solvent Refining Process

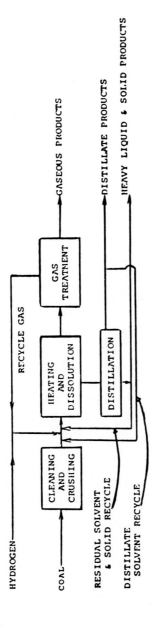

Process Description

The <u>SRC-II Process</u> is very much the same as SRC-I except that distillates are the primary products. Distillation residues which contain unreacted coal and mineral matter (which is believed to be a necessary catalyst) are recycled to the dissolver.

Figure 10-1(b). Solvent Refining Process

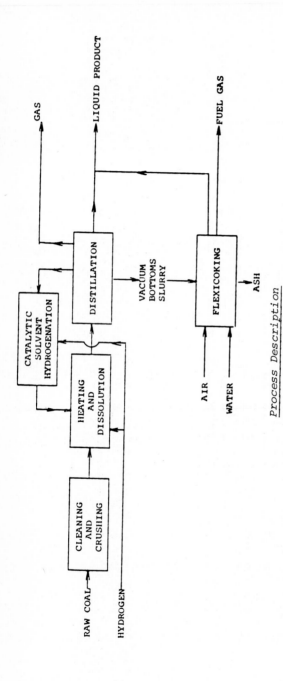

Process Description

Ni-Mo on alumina catalyst, solvent hydrogenation only. Ground coal and hydrogenated recycle solvent are slurried and heated under hydrogen pressure at 800–850°F, 2000 psig operating pressure. Outside catalytic hydrogenation of recycle solvent. Principal product is distillate liquid fuel.

Figure 10-2. Exxon Donor Solvent Process

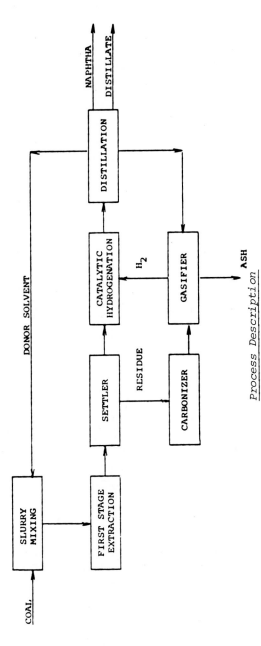

Process Description

In this two-stage process for coal liquefaction, an initial dissolution step is followed, after removal of most of the solids, by a more severe hydrogenation to break down the large molecules and remove heteroatoms. In the first-stage extraction, conditions are relatively mild -- about 300–500 psi, 700°F, with no catalyst or gaseous hydrogen. Recycle solvent and donor solvent are used to achieve the initial dissolution or extraction. The depth of extraction is purposely limited so that subsequent gasification of unconverted residue will satisfy hydrogen requirements. After extraction, a rough separation of solids is achieved by settling to remove coal mineral matter which can foul the hydrogenation catalyst. The settler overhead, containing about one percent solids, then goes to the second stage which is a high severity catalytic ebullated bed hydrogenation and a cobalt-moly catalyst to produce high quality gasoline.

Figure 10-3. CONOCO Two-Stage Liquefaction (CSF)

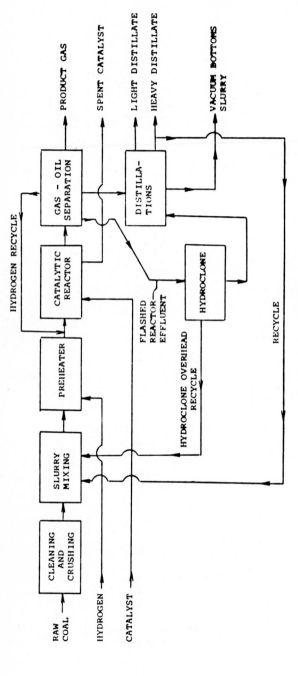

Process Description

Co-Mo on alumina catalyst, ebullated bed. Preheated coal-solvent slurry, under hydrogen pressure is passed through an ebullating bed, at 850°F and 2600-2800 psi operating pressure. Product recovered by vaporization. Ash filtered from bottoms or bottoms are burned. Principal product is "syn crude" or fuel oil, depending on operating conditions.

Figure 10-4. H-Coal Process

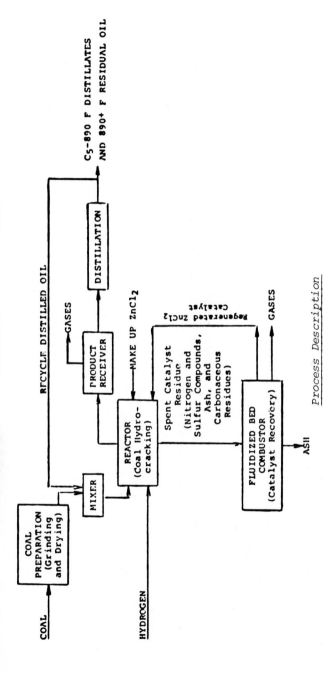

Process Description

This process maximizes the conversion of coal into light distillates by severe catalytic hydro-cracking. Coal or coal extract or SRC blended with recycle oil is fed to the hydrocracking re-actors separately from the molten salt (1/1). H_2 passes up through the reactors at 1500-4500 psig and 650-825°F. Light hydrocarbons and gases are separated from the liquid either in the reactors or in a separate unit and constitute hydrogen recycle, fuel gas, gasoline-range naphtha and recycle oil. In an alternative "catalyst separation" a liquid-liquid extraction column is used in which the effluent catalyst melt is contacted with recycle oil under hydrogen pressure at a lower temperature (600°F) to remove heavy hydrocarbons. The heavy fuel contains 0.3% sul-fur and 0.1% nitrogen.

Figure 10-5. Zinc Halide Coal Liquefaction Process (CONOCO-SHELL)

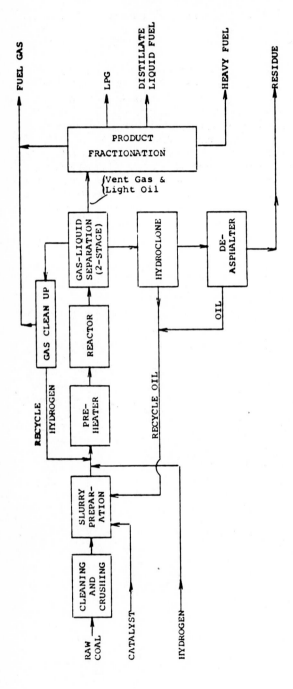

Process Description

Emulsified expendable catalyst. The coal-catalyst-oil slurry and the hydrogen gas is preheated to 660°F. The temperature is raised to 840°F in the reactor. The product then flows to a high pressure separator to remove the hydrogen and light gases. The bottoms then go through a lower pressure separation to remove LPG, light oil, etc. The underflow then goes to the solid separation system. The recycle oil contains catalyst concentrations almost equal to the reactor.

Figure 10-6. Dow Coal Liquefaction Process

Figures 10-1 through 10-6 show simplified block diagrams and the legends give very brief descriptions of the unique features of the processes. Auxiliary processes such as hydrogen and/or oxygen manufacture are omitted.

All of the processes described above have the potential for the production of solid-free liquid fuels from coal. Of those in an advanced stage of development, some, such as the Exxon Donor Process, the SRC-II Process, and the H-Coal Process in the syn-crude mode produce primarily low-sulfur distillates as products. The H-Coal Process in the fuel oil mode also produces a residual material, and the SRC-I Process produces primarily a heavier pro-duct with only a moderate sulfur reduction. The CSF Process is no longer being developed but is included here as an example of process optimization to conserve hydrogen. In its advanced stage (20 TPD), it produced primarily high quality gasoline. A general scheme for all coal liquefaction processes is shown below.

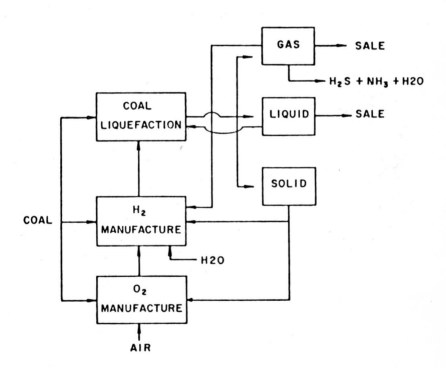

Table 10-1 compares typical yields for each liquefaction
process. All yields are based on Illinois No. 6 bituminous coal
except for the SRC-II process which is based on a Western Kentucky
bituminous coal. These yields do not account for any coal or
residual products going to hydrogen production or to satisfy
energy requirements. The published yields were adjusted to an MAF
basis in each case. The H-Coal (10-4) and SRC-II (10-5) processes
give the lowest residual yield, but accomplish this at a relatively high yield of hydrocarbon gases and a high hydrogen consumption.
Detailed product distributions and compositions for the EDS process under known conditions have not yet been published. The data
given are from the Illinois No. 6 "Base Case" (10-6) before the
flexicoking step (the flexicoker increases the total liquid yield
but decreases the overall product quality since the coker liquid
is poorer than the product from the liquefaction stage). The
overall process appears to produce relatively low yields of a very
high quality and a low quality product with intermediate hydrocarbon gas yield and hydrogen consumption. The SRC-I (10-7) process gives a much higher residual yield with less gaseous products
and lower hydrogen consumption. By running the SRC process at
short contact time (SCT-SRC) even higher yields (10-8) of residual
products and lower yields of gas are obtained. However, the product has about 2 wt % sulfur and an additional upgrading step is
needed to produce a premium fuel. SCT-SRC will be elaborated upon
later in this chapter.

The yields for the SCT-SRC plus upgrading were estimated from
Wilsonville SCT yields and the hydroprocessing of this SCT-SRC by
Mobil (10-9). The actual integration of the SRC production and
upgrading steps might involve recycling of the solvent range product from the upgrading step back to the dissolver. No attempt
was made to estimate the effect of this integration. In addition,
the severity of the upgrading step could be varied to give a
considerably different product slate if desired. The estimated
yields for the SCT-SRC plus upgrading show less gas make than the
other processes and less hydrogen consumption than SRC-II or H-
Coal.

The yields in Table 10-1 are based on MAF coal fed to the
liquefaction reaction. The SRC-II process is run such that the
unreacted coal plus $850^{\circ}F^{+}$ product could be sent to a gasifier to
produce sufficient hydrogen for the process. Accordingly, for
this process about 5.1 tons of vacuum bottoms and/or coal are
required to produce each ton of hydrogen consumed. Using this
number as the basis, the liquid yields from the other processes
were adjusted to reflect the yield based on MAF coal fed to the
liquefaction and gasification sections. This was done as follows:
Assuming 75 percent thermal efficiency, the following requirements to produce one ton of H_2 were calculated: Five tons of

TABLE 10-1. Yield Structure and H-Consumption in Developing Processes

Process	H-Coal Syncrude	SRC-II	SRC-I	SCT[e]-SRC	SCT-SRC + Upgrading	Exxon Donor Solvent
Yields, Wt % MAF						
Hydrocarbon Gases, C_1-C_4	11.8[a]	18.4	8.2[b]	1.3[b]	5.5	5.9[a]
$H_2S + NH_3 + CO_2 + H_2O + CO$	14.1	11.8	9.9	5.2	8.7	15.7
H_2	-6.6	-5.2	-2.9	-1.6	-4.0	-3.9
Distillate Products						
Approximate Boiling Range	C_4-975°F	C_5-900°F	C_6-700°F	C_6-700°F	C_5-650°F	C_4-1000°F
Yield, Wt % MAF	58.4	48.5	27.4	10.1	31.5	34.7
Residual Products						
Approximate Boiling Range	975°F+	900°F+	700°F+	700°F+	650°F+	1000°F+
Yield, Wt % MAF	16.6[d]	22.4[d]	52.4	75.8	49.1	47.5[c]
Unreacted Coal	5.7	4.1	5.0	9.2	9.2	–

[a] C_1-C_3 only.

[b] C_1-C_5.

[c] Feed to flexicoker (which gives additional liquid of lower quality; Exxon has indicated that including the flexicoker the total C_4-1000 F is ~45%).

[d] Ultimately used for hydrogen generation.

[e] SCT = Short Contact Time.

still bottoms, or 5.7 tons of unconverted MAF coal that had passed
through the liquefaction reactor, or 5.8 tons of as-received MAF
coal. For each process, it was assumed that H_2 would be obtained
from these sources in the following order: unconverted coal, then
still bottoms (except in the processes in which this is a
principal product) and then fresh coal. Table 10-2 gives the
liquid yields on this basis along with the elemental composition
and approximate heating value of the coal products. The
composition of the liquid products from the H-Coal, EDS and SRC-II
processes are from the literature (10-10), (10-6), (10-11), while
the SRC-I and SCT-SRC compositions were estimated from the pilot
plant yields of liquid and solid products and our analyses of
these products (10-9). The heating values were estimated from a
correlation based on elemental composition.

Table 10-2 shows that the liquid products from the H-coal and
EDS processes have the highest hydrogen content and heating value,
while the SRC products have the lowest. The liquid yields are
significantly higher for the processes which include the residual
liquids in the product. The SCT-SRC plus upgrading appears to be a
good process for giving a high yield (70.4 wt %) of a high value
liquid product. In addition, higher quality products or higher
liquid yields could be achieved by changing the severity of the
upgrading step. Further work is needed to confirm this. If the
values for hydrogen consumption shown in Table 10-1 are corrected
to grams H needed for 100 g of liquid product, the values range
from 8-12% hydrogen consumption as compared to the stoichiometric
requirement of 5.1 g H/100 g liquid yield.

Both thermal and catalytic processes have been considered
here. We have shown that the products differ in hydrogen and
heteroatom content. The carbon-hydrogen skeletal structures are
believed to be very similar in the two cases, but there are
differences due primarily to the hydrogenation and hydrocracking
activity of catalysts and to temperature differences.

By putting together all of our data on the nature of the
products, we have postulated the types of structures one would
find for typical products containing 35 carbons in an SRC made at
800°F or in an H-Coal product made at 850°F. The results are
shown in Figure 10-7 for molecules that would be found in the
heterocyclic fraction (SESC-3).

The SRC product has fused aromatic rings, very few aromatic
methyls, and more condensed aliphatic rings. The catalytic
product contains single aromatic rings, some aromatic methyls, and
less condensed aliphatic rings.

It is striking that one can draw common skeletal structures
for such varying processes which still conform to the general
analytical observations of the products.

TABLE 10-2. *Yield and Composition of Total Useful Liquid Products*

	H-Coal Syncrude	SRC-II	SRC-I	SCT[c]	SCT-SRC HDT	EDS[d]
Liquid Yield, wt %[a]	51.6	48.5	71.5	85.4	70.4	34.7
Liquid Yield, wt %[b]	58.4	48.5	79.8	85.9	80.6	34.7
Approximate Boiling Range	C_4–975°F	100–900°F	C_5^+	C_5^+	C_5^+	C_4–1000°F
Hydrogen	9.7	8.8	6.3	6.3	8.7	10.5
Sulfur	0.3	0.3	0.6	2.0	0.2	0.8
Nitrogen	0.9	0.8	1.3	1.5	0.7	0.4
Oxygen	–	3.9	4.1	4.9	1.6	2.4
Estimated Heating Value, BTU/lb	18080	17230	16360	15970	17600	18021

[a] Based on MAF coal fed to liquefaction and gasification reactors.

[b] Based on MAF coal fed to liquefaction reactor only.

[c] STC = Short Contact Time.

[d] Exclusive of low quality product from coker.

SRC Conditions (800°F)
Synthetic Solvent (40% Tetralin)

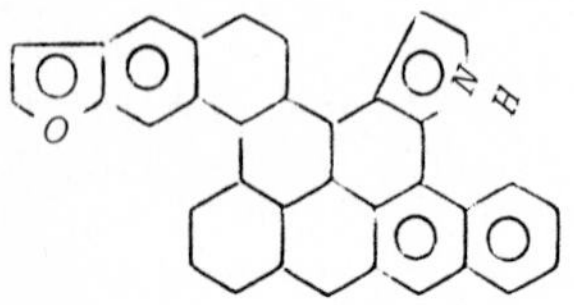

Wt. Avg. Molecular Weight = 483
            % Aromatic C = 63
            % Aromatic H = 39

H-Coal Conditions (850°F)
H-Coal Solvent

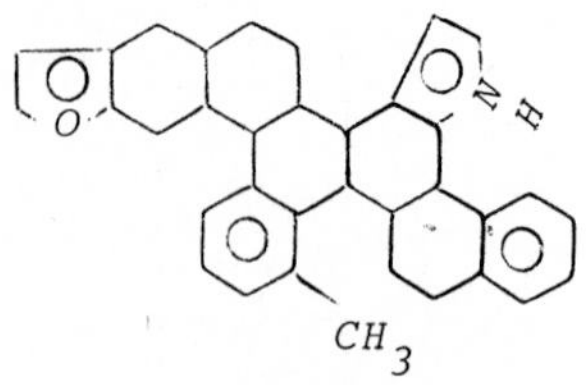

Wt. Avg. Molecular Weight = 487
            % Aromatic C = 57
            % Aromatic H = 32

	%	Empirical Formula	Aromatic	Aliphatic		%	Empirical Formula	Aromatic	Aliphatic
C	86.35	35	22	13	C	86.2	35	20	15
H	6.32	31	12	19	H	7.1	35	12	23
O	4.64				O	3.4			
N	0.95	2			N	1.6	2		
S	1.81	(2.0)			S	0.3	(1.6)		

Figure 10-7. Postulations of the Average Structures of
Heterocyclic Fractions of 800⁺°F Products
(SESC-3) under SRC and H-Coal Conditions.

     The differences in chemistry between the products is con-
sistent with these structures and enables one to envision what
transformations occur via different processes. One can see how
the different liquids can undergo various catalytic and thermal
reactions (hydrogenation, dehydrogenation, and hydrocracking) so
that they have different properties (especially aromaticity) but
virtually the same skeletal structure. Furthermore, all the reactions are reversible, including the type of hydrocracking depict-
ed. We have shown in a study of solvent condensation reactions
that aromatic methyl groups can be incorporated into rings (see
Chapter 9).

III. BASIC STOICHIOMETRIC CONSTRAINTS

Ultimately, all liquefaction processes can be reduced to the
common schematic shown earlier. Coal is dissolved in the presence
of H_2 gas and a recycle solvent to produce gaseous, liquid and
solid products. The required H_2 is obtained from conversion of
some gaseous and/or solid products or perhaps of some feed coal.
Additional feed coal or solids must be burned to provide heat and
electricity for the manufacture of O_2 necessary for the H_2 production. The gas (which must first be separated from H_2, H_2S,
NH_3, and H_2O) and liquid can be sold as products or further upgraded. In an integrated process, heat and H_2 (and the O_2 required to produce it) for further upgrading will require the ex-
penditure of additional coal in some way. At its simplest, any
such scheme reduces to the reaction of air plus water plus coal to
produce saleable products.

Disregarding for the moment such aspects as capital invest-
ment, reliability, operating costs, etc., the success of a coal
liquefaction process and the assessment of its relative merits
hinge on energy efficiency. How much of the energy inherent in
the coal can be sold as useful fuels? For liquefaction, the key
to this is efficiency of hydrogen utilization, which in turn is
chiefly dependent upon the maximization of liquid yields and the
minimization of gaseous hydrocarbon products. As the ratio of
liquid to gaseous products increases, not only does the yield of
liquids from the coal fed to the dissolver increase, but the
amount of additional coal that must be used for O_2 and then H_2
manufacture decreases as well.

It should also be kept in mind that although gaseous hydrocarbon byproducts can be used to regenerate some hydrogen the
overall efficiency suffers, as a portion of the gas must always
be sacrificed due to thermal inefficiencies. Overall thermal
efficiencies of the developing processes for gasification of
solids and liquids range from 60-80 percent. For hydrogen manufacture from light hydrocarbon gases the efficiencies are somewhat
higher (80%).

Liquid product quality is also important. Oxygen, sulfur,
nitrogen, Conradson carbon, metals, ash, and viscosity for instance should be as low as possible, and hydrogen content should
be maximized. These improvements should be effected with a minimum of hydrogen "wasted". Ideally only the hydrogen stoichio-
metrically required in the rejection of H_2S, NH_3, and H_2O and
raising the H/C mole ratio of the liquid product should be ex-
pended. Light gases, although valuable when pure, are produced
in coal liquefaction processes as much less valuable dilute mixtures in H_2. They must either be used or sold as a low BTU gas
or separated at significant expense.

Determination of the actual amount of hydrogen required to convert a given coal to a specific product is not a simple stoichiometric calculation. The major complications are subtle and are the result of the coproduction of certain byproducts as illustrated below.

$$C_a H_b O_c N_d S_e$$

(Coal Feed)

CH_4 (and other light hydrocarbons)

$CO_2 + CO$

$C_{a'} H_{b'} O_{c'} N_{d'} S_{e'}$ (desired product)

$H_2O + H_2S + NH_3$

Low rank coals having high concentrations of carboxylic groups and quinones produce much more CO_2 than higher rank coals. As a result, less hydrogen is required to produce a given H/C mole ratio product than would be predicted from a straightforward stoichiometric calculation. The coproduction of methane increases the hydrogen requirement and will be a function of the particular process.

Correlations have been developed (10-12) which predict the yields of CO and CO_2 during liquefaction as a function of coal rank. These are used in this report to illustrate the sensitivity of hydrogen consumption to coal composition.

Future commercial liquefaction plants will have to be hydrogen self-sufficient. Thus, coal or one of its products will be used to generate the needed hydrogen. The preferable source of hydrogen will undoubtedly be coal liquefaction residues. But, for the sake of simplicity of calculation, we will assume for the immediate discussion that coal is converted quantitatively to the desired fuel, CO, CO_2 and methane, and that the hydrogen required for this transformation is produced from some additional feed coal. (For bituminous coals this calculation procedure is essentially equivalent to hydrogen generation from residue.) Hydrogen generation from coal or heavy liquids can be brought about at 60-80% thermal efficiency depending on the specific gasification process. In the following discussion, we will assume 70% thermal efficiency for hydrogen manufacture. With these assumptions, one can calculate the percentage of feed coal that must be sacrificed for hydrogen manufacture for a series of liquid fuels which are presently of interest. The results are shown in Figure 10-8(a). (All data shown in this figure assume 12.5% byproduct methane alone, as representative of light hydrocarbon gases in general, for simplicity of calculation.) As the quality of the desired product increases from 7% to 13% H, the fraction of bituminous coal needed for hydrogen manufacture increases from 15% to 35%.

These data are, of course, affected by the efficiency of the utilization of hydrogen. The more gaseous byproducts that are produced, the more coal will be needed to supply hydrogen. This point is illustrated in Figure 10-8(b) where we show the fraction of bituminous coal required for hydrogen manufacture at 0% and 12.5% methane byproduct. It can be seen that from 5 to 10% of the coal could be conserved if a more selective process could be developed which eliminates hydrocarbon gas (methane) byproduct formation.

IV. FACTORS TO BE CONSIDERED IN NEW OR IMPROVED COAL LIQUEFACTION TECHNOLOGY.

Up to this point we have reviewed the coal liquefaction processes that have been or are under development and discussed their major uncertainties and limitations. The following statements form the basis for our suggestions concerning new or improved coal liquefaction technology and the research that will be required to develop it.

● The key to success of a process is hydrogen utilization efficiency.

● Solvents can be overhydrogenated.

● The optimal conditions and intrinsic selectivities for catalytic and thermal reactions are different.

● High conversion of coal to soluble products in the initial dissolution step is not necessary when hydrogen-rich liquids are the final product.

● Solids separation presents problems in the processes presently under development.

● Temperatures are very high and reaction times are very long in these processes.

● Denitrogenation is the most problematic and least understood reaction.

Efficient hydrogen utilization is very critical to the economic success of a process and high conversion in the dissolution step is not needed to produce hydrogen-rich liquids from coal. Assuming that all the hydrogen used in a process ultimately comes from the coal (inherent hydrogen plus that

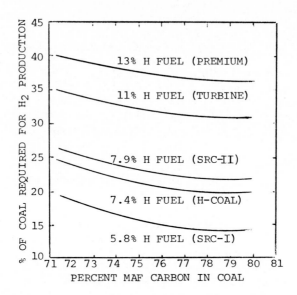

Figure 10-8(a). Hydrogen Requirement for Conversion of Coal

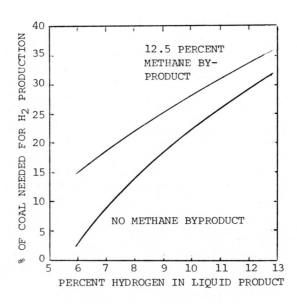

Figure 10-8(b).   Coal Required for Hydrogen Production

obtained from partial oxidation or steam reforming), there is a maximum possible liquid yield dictated by the process thermal efficiency and the elemental analyses and structures of the coal and products.

Solvents can be overhydrogenated if hydroaromatics are further converted to polycyclic paraffins. The latter are neither hydrogen donors nor satisfactory physical solvents for the polar and functional coal products.

Solids separation by filtration presents difficulties because of high viscosities and very small particle sizes. Consequently filtrations must be carried out at high temperatures and filters plug frequently. Distillation is not efficient for solids separation; because of the high solids contents significant amounts of liquids must remain in the distillation bottoms to keep them pumpable.

Thermal reactions require high temperatures because of the relatively high strengths ($\sim$50 kcal/mole) of the bonds that must be broken. However, at such temperatures there are many undesirable side reactions such as cracking, char formation, and dehydrogenation. Aromatics are thermodynamically favored over hydroaromatics at high temperatures. Long times favor these undesirable reactions.

Catalytic reactions have favorable selectivities at lower temperatures where there is an appropriate balance of heteroatom removal, hydrogenation, and molecular weight reduction without excessive hydrocracking to produce light hydrocarbon gases. Also, at lower temperatures poisoning and deactivation via coke formation on catalysts may be less serious. The reactions promoted by catalysts at low temperature are more desirable than the thermal reactions; at higher temperatures a catalyst's selectivity advantage is decreased because of excessive hydrocracking.

Unfortunately, with the present commercial catalysts, nitrogen removal requires both hydrogenation and hydrocracking activities (10-13). Hydrogenation is not undesirable if a hydrogen-rich fuel is the end product, but hydrocracking produces light hydrocarbon gases from other molecules present in the feed. Furthermore, the activity of present catalysts for other reactions such as desulfurization and deoxygenation is reduced by the presence of nitrogen (10-13). Nitrogen-resistant catalysts and denitrogenation with minimal hydrocracking would clearly be desirable.

The conclusion to be drawn from these considerations, is that thermal and catalytic reactions should be decoupled, i.e., run separately under different conditions. We propose that a significant advance in this direction consists of the coupling of short residence time thermal dissolution of coal with catalytic upgrading of the dissolved product, as shown by the data in Tables

10-1 and 10-2. In this way the conditions for each could be optimized, to avoid the poor selectivities of compromise conditions.
It has been shown (10-9c) that the SRC product of short-contact-
time thermal dissolution can be catalytically upgraded in a subsequent step under milder conditions to produce a high yield of
hydrogen-rich liquids with a low hydrogen consumption.

Since a portion of the coal must always be used for hydrogen
production, it is possible to calculate the maximum conversion
that should be attempted in view of the need for hydrogen manufacture. A similar procedure was used to optimize the utilization
of the coal in the CFS process (10-14). Consider, for example,
the production of turbine fuel having 11% hydrogen. Bituminous
coals need only be converted to an extent of 69%, subbituminous
coals to about 67%, and lignites to about 65%

We discussed in Chapter 5 the fact that conversions in this
range are possible with short-contact-time thermal coal dissolu-
tions with solvents similar to SRC recycle solvents under conditions equivalent to those found in preheaters. This is illustrated in Figure 10-9.

The conversions shown in Figure 10-9 were obtained at 800°F.
If higher conversions are desired, higher temperatures could be
used or the solvent could be enriched with additional hydrogen
donors (Chapter 9). Optimal solvent properties for this step are
still not completely understood and more research in this area is
warranted.

The major problems associated with such a process where short
contact time SRC (SCT-SRC) and solvent are taken directly to a
severe catalytic step, would be in maintaining suitable solvent
quality and catalyst activity, and in solids removal.

The solvent-range material produced in the upgrading stage of
such a process would not be expected to be a good solvent in the
first step, because it would have been overhydrogenated (aromatics
would have been hydrogenated past the hydroaromatic stage). Such
a solvent would not be a good physical solvent because the aromatics content would have been reduced and such functional groups
as phenols removed.

The short contact time dissolution step as described so far
probably would not be solvent self-sufficient because insufficient
solvent-range material would be produced from coal to offset solvent losses due primarily to condensation reactions. Also, solvent-range material exiting this reactor would probably contain
negligible donors.

Thus, separate recycle solvents for each stage of the conversion appear optimal. Removal of solids between stages could
be difficult because of high liquid viscosity and functionality.

The first stage effluent, even after solids removal, would not be a good feedstock for a catalytic process because it would contain large amounts of coke-precursors and metal contaminants, both of which would contribute to rapid catalyst deactivation. A procedure must be developed to (a) generate a suitable solvent for the dissolution step, and (b) upgrade the feedstock, prior to treatment over the conventional catalyst, enough to protect that catalyst from rapid deactivation.

At the present time there are two potentially viable solutions to the problems associated with using short contact time coal dissolution for the production of a solid boiler fuel that can be catalytically upgraded to premium liquid products of higher hydrogen content.

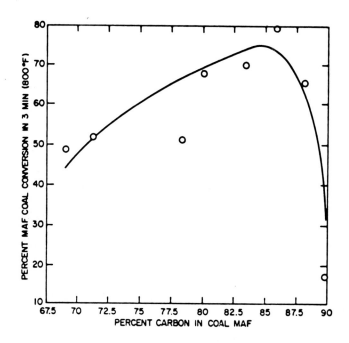

Figure 10-9.   Short-Time Coal Conversion vs. % MAF Carbon

The first of these is a process variation known as SRC-½ now under preliminary development. A simplified schematic and explanation are shown in Figure 10-10. There are two key features of this scheme that would allow short contact time conversions to be run efficiently and with sufficient generation of a suitable solvent. First, the solids removal is done via the Kerr-McGee Critical Solvent Deashing (CSD) process. In the CSD process, SRC is separated from recycle solvent by distillation and then contacted with a second, deashing, solvent. At approximately the critical conditions of this latter solvent, small changes in the temperature and pressure result in major changes in SRC solubility. The separation is run in three stages. At the first set of conditions, unreacted coal and mineral matter are insoluble and can be readily removed by settling. The conditions are changed and the bulk of the SRC becomes insoluble and is removed as the product of the process. The conditions are changed again and the remaining SRC can be separated (10-15).

This last fraction is known as light SRC. Recycle of the light SRC to the short contact time reaction is the second major feature of the SRC-½ process. The total amount of recycled material is increased, making the process solvent-balanced. Furthermore, and most important, the light SRC appears to have remarkable solvent powers, greatly increasing the extent to which the coal and initial dissolution products can be hydrogenated in the short contact time reactor.

The second promising solution to a practical process for short contact time dissolution plus catalytic upgrading is in the use of inexpensive disposable or regenerable catalysts. These could be used to effect a small but necessary increase in the yield and hydrogen donor ability of recycle solvents in SCT processes, to upgrade the SCT-SRC products somewhat, and to remove metal contaminants and the worst coke precursors in order to protect a subsequently used more expensive catalyst.

There are two attractive sources of such disposable catalysts. The first is coal mineral matter itself, in particular pyrite; the effects of such materials were described in Chapter 6. In fact, the SRC-II process is essentially a scheme designed to increase the mineral matter/coal ratio. Another way to use coal mineral matter would be to clean a coal for use as a fuel, for instance by float-sink or magnetic techniques, and then use the mineral matter obtained to aid in the liquefaction either of more of the same coal or of a different coal.

The second source of disposable catalysts is naturally-occurring minerals. An example is red mud, reported primarily by Japanese workers. This material, primarily oxides of silicon, aluminum and iron, is the residue from aluminum recovery from bauxite. Another example is manganese nodules, ores containing a

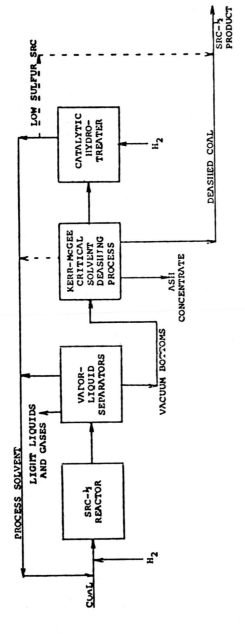

Process Description

The essential element of this process is a SRT reactor, the product slurry from which is dis-
tilled to produce a middle distillate and a vacuum bottoms. The bottoms are separated by Kerr-
McGee critical solvent deashing (CSD) into three streams. The lightest material (polymerized
solvent and light SRC is catalytically hydrogenated to produce a recycle solvent component and
a low sulfur component to be blended with the SRT-SRC. This blend constitutes the ultimate SRC
product. Ash and solid residue are rejected as the third CSD product.

Figure 10-10. SRC-½ Conceptual Process Diagram

variety of metals, especially iron, found on lake and sea beds.
These porous, high surface area materials could be used in coal
processing and still retain their value as metal ores. Catalytic
activity for a variety of reactions appropriate to this envisioned
use has been reported (10-13).

Disposable catalysts could be used in the first stage of a
short contact time process by co-feeding them with the coal; high
concentrations could be allowed to build up in the reactor if
desired. Alternatively, a second reactor could follow the short
contact time reactor. Either technique could be incorporated into
the SRC-$\frac{1}{2}$ scheme. This combination could be optimal.

Short contact time dissolution, followed by separate
catalytic upgrading to make higher quality products if desired,
appears to be the most efficient way to make clean fuels from coal
in light of our current understanding of the chemistry and
technology of coal liquefaction and the reactions of coal-derived
materials.

REFERENCES

10- 1. M. Berthelot, Bull. Soc. Chim. Fr., 11, 278 (1869)

10- 2. F. Bergius, British Patent 18232 (August 1, 1914).

10- 3. A. Baba, Chem. Econ. Eng. Rev., 7, 15 (1975).

10- 4. H-Coal Integrated Pilot Plant, Report Number L-12-CL-510, Volume 1, July 1977.

10- 5. B. K. Schmid and D. M. Jackson, The SRC-II Process, Third Annual International Conference on Coal Gasification and Liquefaction, Univ. of Pittsburgh, August 3-5, 1976.

10- 6. Exxon Donor Solvent Coal Liquefaction Process Development Phase III-A. Final Technical Progress Report for the period January 1, 1976-June 30, 1977, Volume 1, Exxon to DOE & EPRI, February 1978.

10- 7. "Solvent Refined Coal Pilot Plant, Wilsonville, Alabama", Technical Report No. 8, Catalytic Incorporated to Southern Services, April 30, 1976.

10- 8. "Operation of Solvent Refined Coal Pilot Plant at Wilsonville, Alabama", Annual Technical Progress Report, January-December 1976.

10- 9. (a) T. R. Stein, R. H. Heck, et al., "Upgrading of Coal Liquids for Use as Power Generation Fuels", Annual Report for the period February 1977-January 1978 from Mobil Research and Development Corp. to the Electric Power Research Institute, Report AF-873, December 1978.
(b) ibid., EPRI Report AF-444, October 1977.
(c) ibid., Proceedings of EPRI Contractors' Conference on Coal Liquefaction, May 1978.

10-10. D.D. Whitehurst, M. Farcasiu, T.O. Mitchell, and J.J. Dickert, Jr., "The Nature and Origin of Asphaltenes in Processed Coals", EPRI Report AF-480, Second Annual Report under project RP-410-1, July 1977.

10-11. "H-Coal Process Passes PDU Test", Oil and Gas Journal, pp. 52-53, August 30, 1976.

10-12. D. D. Whitehurst, unpublished results based on data in C. H. Fisher, G. C. Sprunk, A. Eisner, H. J. O'Donnell, L. Clarke, and H. H. Storch, U. S. Bureau of Mines Tech. Paper 642, Washington, D. D., 1942.

10-13. D. D. Whitehurst, T. O. Mitchell, M. Farcasiu, and J. J. Dickert, Jr., "Exploratory Studies in Catalytic Coal Liquefaction", Final Report from Mobil Research and Development Corp. to EPRI under project RP-779-18, AF-1184, September 1979.

10-14. Fluor Engineers and Constructor's Report FE-2251-40, "Conceptual Design for Advanced Coal Liquefaction Commercial Plant", October 1977 (also references cited therein).

10-15. R. M. Adams, A. H. Knebel, and D. E. Rhodes, Chem. Eng. Prog., June 1979, p. 45.

Index

A

Aromatics, content in solvent-refined coal, 39–42, 193, 311

B

Benzophenone, 289, 290
Bond-breaking, during coal dissolution 287, 300

C

Carbonization, 207, 210, 217, 218, 222, 224, 228, 231, 237
  SESC fractions, 219–245
Catalytic upgrading of solvent refined coal, 163, 169, 172, 357–360, 365–369
  cobalt–molybdenum catalyst, 163, 165, 168
  reactions
    dehydroxylation, 162, 164, 165, 168, 169
    deoxygenation, 76–80
    desulfurization, 76–80
    hydrocracking, 170
    hydrogen transfer, 167, 168
    hydrogenation–dehydrogenation, 162–165, 168, 174
    methyl transfer, 168
    role of hydrogen gas, 162, 163, 166, 167, 172–174
  sulfur requirement, 168
Char
  compounds forming, 289–333
  hydrogen source, 335
  type
    anisotropic carbon, 207–211, 213, 216, 217, 222, 237–247, 253, 258
    cenosphere, 258, 260
    coalescence, 216, 228
    domains, 208, 209, 213, 216, 243
    flow-type, 209, 211, 213, 214, 247

    mesophase, 207, 208, 211–218, 228–237, 243–245, 253, 258, 262, 266
    mosaic, 208, 209, 213, 246, 248
    glassy carbon, 207, 208
    isotropic carbon, 207, 208, 217, 223, 245, 249, 257, 260, 262, 266
    needle-coke, 213, 214, 217, 218
Char formation, see also Regressive reactions
  catalysis of, 209, 212, 213
  due to hydrogen acceptors, 333
  effect of partial hydrogenation, 214, 218
  experimental methods
    autoclave, 249–266
    in pressurized gold tubes, 224–248
    thermogravimetric analysis, 219–223
  from heavy phenols, 337
  mechanisms, 207–218, 226, 227, 266, 271
    effect of mineral matter, 216
    inhibition by H-donors, 224, 230–232, 237–242, 248, 258, 260, 264, 270, 271
    promotion by H-acceptors, 231, 232, 239, 242, 262, 264, 269
  from solvent, 334–336
  with reaction time, 137, 138, 297
    effect of process variables, 178, 195, 198–201, 204–205
Coal composition, see also specific coal
  aromaticity, 9–11, 19, 25, 193
  change with
    heat treatment, 115, 116, 118–120
    oxidation, 113, 114, 115, 119, 120
  as a function of rank, 146, 147
  functional groups, 13–20
    oxygen-containing, 13–19, 21, 22
      carbonyls, 13, 16–18
      carboxylic acids, 13, 16–18
      change during liquefaction, 76–80, 131–144, 149, 152–154, 157

S